MW00423727

Marshal Maurice de Saxe, 1696-1750
From An Old Drawing

REVERIES ON THE ART OF WAR

Maurice de Saxe

Translated and Edited by
Brig. General Thomas R. Phillips

Dover Publications, Inc.
Mineola, New York

Bibliographical Note

This Dover edition, first published in 2007, is an unabridged republication of the edition published by The Military Service Publishing Company, Harrisburg, Pennsylvania, 1944. The work was originally published in London in 1757.

Library of Congress Cataloging-in-Publication Data

Saxe, Maurice, comte de, 1696–1750.
[Mes rêveries. English]
Reveries on the art of war / Maurice de Saxe ; translated and edited by Brig. General Thomas R. Phillips.
p. cm.
Previously published: Harrisburg, Pa. : Military Service Pub. Co., 1944.
ISBN 978-0-486-46150-2
ISBN 0-486-46150-5
1. Military art and science—Early works to 1800. I. Phillips, Thomas Raphael, 1892– ed. and tr. II. Title.

U101.S273 2007
355.02—dc22

2007004582

Manufactured in the United States of America
Dover Publications, Inc., 31 East 2nd Street, Mineola, N.Y. 11501

CONTENTS

TO THE READER

I wrote this book in thirteen nights: I was sick; thus it very probably shows the effects of the fever I had. This should supply my excuses for the irregularity of the arrangement, as well as for the inelegance of the style. I wrote militarily and to dissipate my boredom. Done in the month of December, 1732.

INTRODUCTION

ELDEST of 354 acknowledged illegitimate children of Frederick Augustus, Elector of Saxony and King of Poland, Maurice of Saxony, the Prodigious Marshal, was born October 28, 1696. His mother was the lovely Countess Aurora von Konigsmark. Frederick Augustus was renowned for his fabulous strength, the immensity of his appetites, and his limitless lust. Maurice inherited these characteristics, but combined with them a very superior intelligence.

We are told by General Theron Weaver (*The Military Engineer,* 1931, in his *"Flanders Field–1745"*) that he "strongly resembled his father, in both person and character. Carlyle describes him . . . as having 'circular black eyebrows, eyes glittering bright, partly with animal vivacity, partly with spiritual,' and as standing well over six feet in his stockings. He was very strong, and could bend a horseshoe with one hand. He was further distinguished by being 'the worst speller ever known;' this in a world of bad spellers."

He was tutored at his father's expense until the age of twelve, when he was placed under the tutelage of General von Schulenburg, one of the most distinguished soldiers of fortune of the time. At the same time he was commissioned an ensign in the infantry and marched on foot from Dresden to Flanders where he fought in the battle of Malplaquet [1709; English, French and Dutch defeated French] under Marlborough and Prince Eugene of Savoy, being then thirteen years of age. While the slow operations of eighteenth century sieges were being carried out he found time to seduce a young girl in Tournay, by whom he had a child, thus proving himself a true son of his father.

When the peace of Utrecht was signed Maurice was 17 years old, had made four campaigns in Flanders and Pomerania, had distinguished himself for intrepidity, and commanded his own regiment of horse, which he drilled and fought with single-minded passion. He was married, much against his will, at the age of 18 to an immensely wealthy 14-year-old heiress and started wasting her fortune as rapidly as he knew how. It maintained his regiment of horse and his legion of mistresses. The war against the Turks provided an opportunity for his talents, and he served with Prince Eugene in the capture of Belgrade in 1717. Having squandered the fortune of his wife, the court of Versailles beckoned him as the most likely place for a distinguished soldier to make a new fortune, and the year 1720 found him in Paris.

Mars and the Ladies

He was accepted and lionized in the degenerate society of the French capital and became a bosom friend of the Regent [Duc d'Orleans, between the reigns of Louis XIV and Louis XV]. In August, 1720, he was commissioned Field Marshal in the French army and purchased a regiment. But his greatest successes were with the French ladies of the court. His regiment represented the serious side of his nature, and he trained this with the utmost thoroughness. Between debauches he set himself to studying tactics and fortification and reading the memoirs of great soldiers.

In Paris he became the acknowledged lover of the lovely Adrienne de Lecouvreur [1692-1730], the greatest tragic actress of the age and the toast of France. She was accepted in the noble society of Paris, and Voltaire [philosopher, author, 1694-1778; friend of Frederick the Great] was her sincere friend. Having run through his wife's fortune, his marriage was annulled.

Balked of a Throne

The throne of the Duchy of Courland having become vacant in 1725, Maurice plotted to gain it. He would have to be elected by the Diet. Anna Ivanowa, niece of Peter the Great [of Russia, 1672-1725], likewise had a party favoring her. Saxe's scheme was to marry Anna Ivanowa and combine their claims. But money was needed to finance the plan. There were plenty of women in Paris to supply it, and even Adrienne sold her jewels to help her hero to marry another woman and to gain a throne. But Peter had another idea. This was that Saxe should give up his claims to the Duchy of Courland, marry Elisabeth Petrovna, daughter of Peter the Great and be satisfied with the portion she would receive.

Maurice pleased the Dowager Duchess of Courland better than she did him. While living in her palace, supported by funds supplied by Adrienne and other women, "he was caught out in a most ridiculous affair with one of the Duchess' ladies-in-waiting, which brought his career in Courland to an abrupt conclusion," narrates Thornton in his *Cavaliers, Grave and Gay.*

"One snowy winter's night, Saxe was taking the lady back to her own quarters. By ill luck they ran into an old woman with a lantern. In order to conceal his companion's identity Saxe made a kick at the lantern, slipped up in the snow, and lady and all fell on top of the old woman, whose screams fetched out the guard. The whole matter was reported to the Duchess, who was furious, and although Saxe assured her that it was 'all a dreadful mistake,' she was not to be mollified. The outraged Duchess threw him out of the palace, and out of her life."

Pompadour His Patroness

In the wars between 1733 and 1736, ending in the Peace of Vienna, Maurice again distinguished himself and was made a lieutenant general. Still, he was nothing but a German nobleman and a military adventurer. His fortunes commenced to improve with his acquaintance with Madame de Pompadour, [1721-64], mistress of King Louis XV. She recognized his greatness as a soldier and his defects of character as a man. "Maurice de Saxe," she wrote, "does not understand anything about the delicacy of love. The only pleasure he takes in the society of women can be summed up in the word 'debauchery.' Wherever he goes he drags after him a train of street-walkers. He is only great on the field of battle." [quoted by George R. Preedy, *Child of Chequer'd Fortune*]. It was probably due to her influence with King Louis XV that Maurice was retained in high rank in the French army and was given supreme command, in spite of the claims and jealousy of the princes of the blood.

In the great war of 1740 [of the Austrian Succession, 1740-48, against the accession to the Austrian throne of Maria Theresa, 1717-80] she • retained her throne, but lost territory under the military force of Spain, France, Prussia and Bavaria], Saxe, then a lieutenant general, was sent with a division of cavalry to the aid of the Duke of Bavaria. On the invasion of Bohemia, he led the vanguard. It was by his advice and under his direction that Prague was attacked and carried.

"It was this exploit," as General Weaver describes it, "which made his name famous throughout Europe as a military commander." He says:

"Here, too, were evidenced the advanced ideas of the man. The assault was made against one of the city walls under brilliant moonlight by a small portion of Saxe's force under his own personal leadership, while two larger forces made attacks at other sides of the town; one, a real attack which failed, and the other a feint. The results of these two attacks were successful, in that the defending forces were drawn away from the side which was assaulted by Saxe. Detailed orders and plans has been made for the assault. Upon reaching the walls, however, it was discovered that the scaling ladders were too short; but Saxe, perceiving a nearby gallows, spliced his long ladders with the short ones leading to the platform [of the walls] and thus succeeded in getting . . . fifty men over the wall. The lone sentry was sabered and the gate then opened for Saxe and the remainder of his men.

"The town was captured with very little bloodshed, and what is most remarkable of all, without looting. Saxe had given strict orders that there was to be no straggling and that if any straggler was discovered he would be bayoneted on the spot; and also that any trooper found off his horse would be sabered."

Rise of Star of Glory

The period of Maurice's glory commenced in 1745 when he was appointed a marshal of France and placed at the head of the army with which Louis XV in person proposed to conquer the Netherlands. His difficulties were greatly augmented by presence of the King and court who came to check rather than assist operations. Notwithstanding these drawbacks, the campaigns of 1745-46-47-48 reflect the greatest credit on his military skill and sagacity. The capture of Ghent, Brussels, and Maestricht, the battles of Lafeldt, Roucous, and Fontenoy, were all splendid feats, and were due to his wise planning and dispositions.

Ill with dropsy and hardly able to move, he left for the campaign that resulted in the victory of Fontenoy [of French over the English and allies, 1745]. He encountered Voltaire, who asked him how could he do anything in his half-dead state. Saxe answered: "It is not a question of living, but of acting."

In planning his celebrated Flanders campaign of 1745, late in the previous year, Saxe, according to Weaver, "made what must go on record as a perfect estimate of the situation as it would, and did, develop, some six months in advance of the execution of his plan."

The armies finally met at Fontenoy, name of a ridge and a village in Belgium. Saxe commanded the French and the Duke of Cumberland the British and their allies. Saxe had about 80,000 men and the Allies all told close to the same number. To meet the expected Allied offense, Saxe had prepared an admirable system of barricades, entrenchments, abatis and three mutually supporting redoubts between the villages of Fontenoy and Antoing. Two more isolated redoubts were constructed. He committed one error, as it later developed, by not providing a sixth redoubt in the center of a 900-yard opening between the

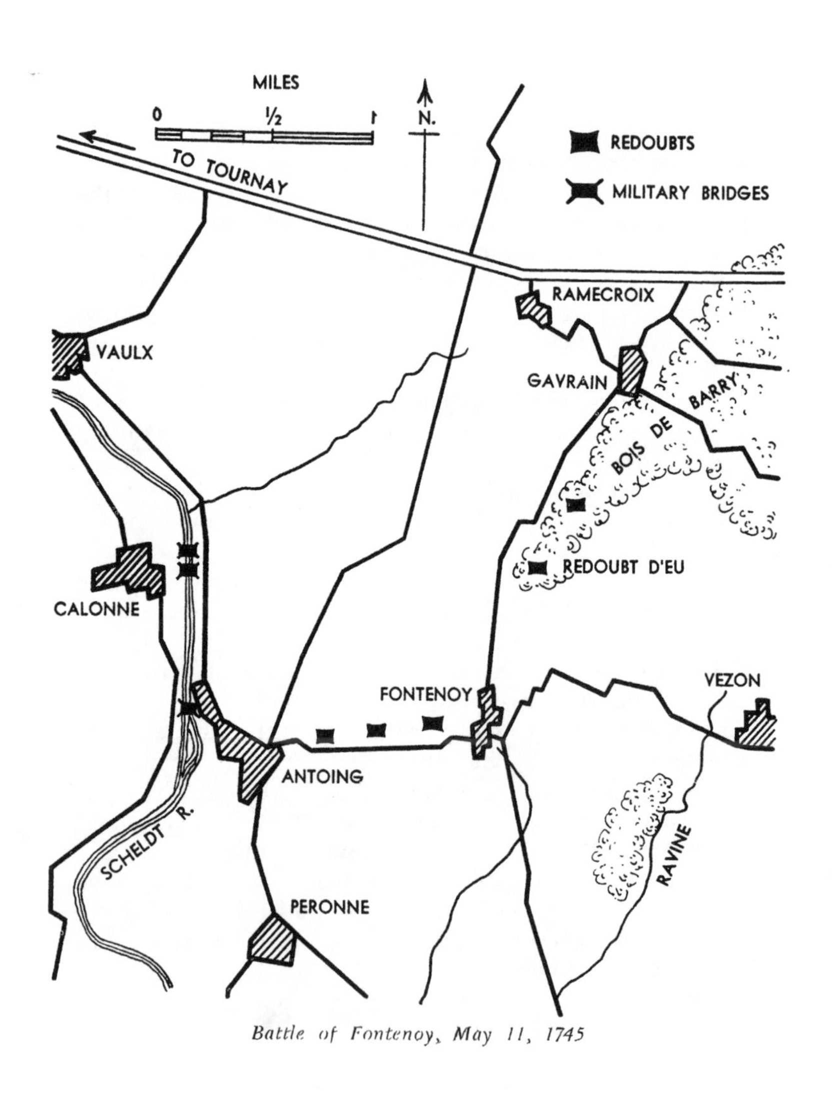

Battle of Fontenoy, May 11, 1745

northern limit of Fontenoy village and his fifth redoubt. He had 100 cannon on his front and a subsidiary battery near Antoing.

A Tough Old Trooper

The Allies attacked on the morning of May 11th. One of the first discharges from French cannon "struck General Sir James Campbell, the cavalry commander, on the knee. General Campbell seems not to have known of our present day compulsory retirement at 64 years of age, for he was then seventy-three and still going strong. His dying remark was that to the effect that his dancing days were over." Weaver goes on:

"On the French side there were two rather remarkable points to be noted. First, Maurice de Saxe was a very sick man, an attack of dropsy made it necessary for him to be carried about on a litter the greater part of the time, and also making him so thirsty that he is said to have chewed on a lead bullet all day in an effort to quench his thirst. The other important and most unusual circumstance was that Louis XV himself was present on the battlefield . . . in a place of comparative safety . . . well in rear of the lines. . . . It was more or less due to the urgings of a former mistress, the Duchess de Chateroux, that Louis had been prodded into a little activity and desire for glory as a king who led his troops in battle.

"The duchess had died and her place in the sun usurped by the more famous Madame de Pompadour, but Louis still retained part of his desire to display himself before the troops; and what was of much greater importance, he had unwavering confidence in Saxe, and a willingness to support him by his kingly presence against the numerous political enemies of the Marshal. It . . . is to the everlasting credit of Louis that on this day at Fontenoy he, for once in his unheroic life, behaved like a real man and a king."

Dealing With the "British Lobster"

"Saxe's enemies had advised the king to withdraw and

had criticized the whole plan of battle. The reply of Saxe to this was: 'So long as I am cook in charge of the stove I intend to deal with the British lobster in my own way.' Louis said to Saxe in the presence of these critics: 'When I chose you to command my army, I intended that you should be obeyed by everyone, and I myself shall be the first to set the example!'

"Flank attacks by the Allies failed. Cumberland decided to risk all by storming the gap between Fontenoy and a wood, where Saxe had not deemed it necessary to put a redoubt, not believing that the enemy would venture to break through on ground that was commanded by the French artillery. But the British did. Between 14,000 and 16,000 infantry, in ranks of six files, behind which were strong cavalry, crowded into the gap, flags flying and drums beating, in parade order. The mass doggedly pressed forward until it gained the crest of the Fontenoy ridge, 800 yards away.

Birth of a Legend

"The Allies halted briefly on the crest, while the front gradually contracted, as the regiments on the flanks pressed inward to avoid the artillery fire. Then the French infantry advanced slightly from their positions on the reverse of the slope, and the French Guards and the British Guards met face to face, with only some 30 paces separating them. It was here that one of those epic incidents of war took place.

"The legend . . . states that at this time Captain Lord Charles Hay (Lieutenant of the First Grenadier Guards) removing his hat and taking a drink from a flask which he drew from his pocket, advanced toward the French and said to the Marquis d'Auteroche, who in wonder stepped forward a bit: 'Monsieur, bid your people fire.' The Marquis is supposed to have replied: 'No, monsieur; we never fire first.' Hay then led his men in a cheer, to which only a few of the French responded. The French then fired, but ineffectively. The British in turn fired one tremendous series of volleys by companies, and 50 officers and 750 men of the foremost French regiments fell at once. Legend and truth are evidently at variance here, for Lord Hay in a letter to his brother stated that he had actually said to Auteroche: 'Gentlemen of the French Guards, I hope you will stand and fight us today and not escape us by swimming the Scheldt, as you did by swimming the Main at Dettingen. [in 1743].'

"The French infantry fled in a panic, running through their cavalry in the rear. The day appeared to be lost for the French as the Allied mass methodically pressed forward. . . . Saxe understood the peril of the situation and in spite of his illness mounted his horse and dashed about the field, collecting every stray man or unit he could lay hold of and throwing them against the British mass, in a manner somewhat similar to that used by the British against the Germans when the latter broke through [in France] in 1918. Luckily for the French, the Dutch and the Austrians on the Allied left did nothing, and even though in twenty minutes the British had overrun the French local reserves and were in the French camp, Saxe, by bringing up troops from Fontenoy, was able to stop the British advance.

Turn of Battle's Tide

"The French courtiers urged the king to ride in safety over the bridges [built by Saxe, in case a retreat was necessary], but Saxe on hearing this, and in the presence of the king, said: 'What damned coward advised the king to retire? All is not yet lost. Fight or die!' . . . The king stayed with the marshal.

"Under French infantry and cavalry attack from all sides, the British halted, drew in flanks and formed three sides of a hollow square. By midday French reserves approached. Artillery was brought up and cut up the square until it finally broke ground and retreated, eight minutes after the guns opened fire on it. The battle was over at two and a half hours after noon.

"Casualties were about even, 7,000 on each side, including more than 4,000 British. The Allies fell back to Lessines, unpursued. 'We had had enough of it,' Saxe acknowledged, 'There was no artillery ammunition left and the cavalry was fought to a standstill.' " With the capture of Brussels in the following February, after Saxe had taken town after town, the Flanders campaign ended.

Rewards and Faded Dreams

The material gains to Saxe were enormous. For his victories Louis made him a count. He revived for him the title and rank of "Marshal General of the King's Camps and Armies," which had formerly been held by Turenne. He gave him the chateau of Chambord, which carried a revenue of between seven and eight million francs, a pension of 40,000 francs and six captured cannon to display in the park of Chambord. Here he had his own regiment of cavalry, barracks for them, and a parade ground large enough for drill. He had his own theater and troupe of actors. With the exception of a visit to Frederick the Great in the summer of 1749, where he was entertained in the royal cottage of Sans Souci near Berlin, in the most lavish manner, Maurice passed the remainder of his days in the society of artists, men of letters, and courtesans. He still evolved various schemes to provide himself a kingdom, including the crown of Corsica, the island of Tobago, and even the founding of a Jewish kingdom in South America. Death put an end to these dreams and closed his career at Chambord, in November, 1750, in the fifty-fourth year of his age. His last words, addressed to his medical attendant, Monsieur de Senac, were: "Life, doctor, is but a dream, and I have had a fine one." Louis XV's comment upon the loss of Saxe was: "I have no generals, now; only captains."

Weaver comments that "some vestige of his literary genius, at least, has been passed down to us, for an illegitimate daughter was the great grandmother of George

Sand. [Amantine L. A. Dudevant, French novelist, 1804-76]."

The *Reveries* were published posthumously in 1757 and were translated into the English the same year. They were variously appreciated. Carlyle notes that Frederick the Great gave a copy of them to the Kaiser Joseph [of Austria] who kept them thenceforth on his night table, where they were found after his death, twenty-one years later, not a page read, the leaves sticking together. Carlyle calls them "a strange Military Farrago, dictated, I should think, under opium." [Probably no less unsound than some of the unrestrained eulogies of Frederick penned by the Scot].

Ahead of His Times

They evince a deeper insight into tactics and leadership than any other work known in Europe, since the Romans to his time. Saxe was not merely far in advance of his age in tactical conceptions and the influence of the human heart on battles, but he was far in advance technically. He wanted breech-loading cannon and muskets. He invented his *"amusette,"* an accompanying gun for the infantry. He wanted to reform the uniforms to make them practical rather than show clothes.

He advocated company messes instead of small group cooking. He planned to break up enemy charges with specially trained groups of expert riflemen and skirmishers, who then would retire and leave the disorganized enemy an easy prey to his counterattack. His advocacy of conscription antedated all previous ideas on this matter. He rediscovered cadenced marching, lost since the time of the Romans, which was to change European armies from straggling mobs into disciplined soldiers. He was the first to object to the practice of volley fire, recognizing that accurate aim became impossible if men were forced to hold their muskets in aiming position indefinitely while awaiting the command to fire.

In the tactical field Saxe expressed his scorn for entrenchments, saying they always were taken. In their place, he would use redoubts, the eighteenth century equivalent of the modern strongpoint. The soldiers of World War I had to go through the same evolution, starting with entrenched lines and ending in mutually supporting strongpoints. He attempted to open up the formations of the day and ridiculed charges in mass, which soon became confused and broke into a useless mass of men who only got in each other's way. Saxe was also the first soldier of modern times to advocate implacable pursuit of a defeated enemy.

Legs Against Arms

A stern disciplinarian, Saxe did not overemphasize the importance of drill. "Drill is necessary to make a soldier steady and skillful, but it does not warrant exclusive attention. Among all the elements of war it even is the one that deserves the least." But marching is another matter. "The foundation of training depends on the legs and not the arms. All the mystery of maneuvers and combats is in the legs, and it is to the legs that we should apply ourselves. Whoever asserts otherwise is but a fool and lacks the elements of what is called the profession of arms."

Saxe also was an innovator concerning organization. He wanted to reorganize armies after the Roman fashion, modernized, with legions and smaller units. In his time armies were divided into three or four wings regardless of size. His scheme was the progenitor of our present division organization. He also would designate organizations by number, instead of by the name of the colonel, as was the custom of the time. In support of this he argued that this would give the organization a continuing history and would result in increased esprit de corps. He also would give each man insignia to denote his regiment.

Saxe's stature is most clearly shown in his appreciation of the moral factors in war and in his conception of leadership. He states concerning Chevalier Follard, that he errs in supposing "all men to be brave at all times, and does not realize that the courage of the troops must be reborn daily, that nothing is so variable, and that the true skill of a general consists in knowing how to guarantee it by his dispositions, his positions, and those traits of genius that characterize great captains . . . It is of all the elements of war the one that is most necessary to study. Without a knowledge of the human heart, one is dependent upon the favor of fortune, which is sometimes very inconstant."

Why are panics aroused? "This is because they are faced with the unexpected and fear for their flanks and rear. In all probability they will take flight without knowing exactly why."

His Ideal General

For the general, Saxe had high ideals. He should possess a talent for improvisation. His plans should be complete and meticulous, his orders short and simple, but on the day of battle he should be occupied with nothing but the conduct of the action. "Thus, on the day of battle, I should want the general to do nothing. Many commanding generals only spend their time on the day of battle making their troops march in a straight line, in seeing that they keep their proper distances, and in running about incessantly themselves. In short, they try to do everything and as a result, do nothing. They appear to me like men with their heads turned, who no longer see anything, and who only are able to do what they have done all their lives, which is to conduct troops methodically under the orders of a commander."

What is the cause of this? "It is because few men occupy themselves with the higher problems of war. They pass their lives drilling troops and believe that this is the only branch of the military art. When they arrive at the command of armies, they are totally ignorant, and in default of knowing what should be done, they do what they know."

Outside of the military terrain, Saxe's book contains numerous ideas far in advance of his time. One curious chapter dealt with "Reflections on the Propagation of the Human Species," which one modern commentator observes appears to be incomplete without "an introductory foreword written by Augustus the Strong, who must have been eminently fitted to give expert testimony on this subject, at least." The chapters on "Morale" and "Human Nature in War" possess value even now, for they set out views and understandings of soldiers minds which modern commanders might profitably consider. The following quotation expresses one of Saxe's ideas:

"Man is an engine, the motive power of which is the soul; and the greatest amount of work will not be accomplished by this engine for pay, or under pressure, or by the help of any kind of fuel which may be supplied by the ton. It will be done only when the will or the spirit of the creature is brought to its greatest strength by its own natural fuel, namely: the affections."

This translation is completely new and has been made from the text of Charles-Lavauzelle published in 1895. Certain portions of no present value have been omitted. These include a long project for an invasion of Poland and a number of details without modern interests. The only previous translation into English was made in 1757 and is so inaccurate as to nullify frequently many of Saxe's most brilliant remarks. In particular the sentences dealing with drill, which are probably the most quoted extracts from Saxe, were so rendered in the old translation as to reverse the sense of the author.

Two novelized biographies of Saxe in English have been published recently: *The Prodigious Marshal,* by Edmund B. D'Auvergne, Dodd Mead & Co., New York, 1931, and *Child of Chequer'd Fortune,* by George R. Preedy, Hubert Jenkins, Ltd., London, 1939. Both are excellent and make fascinating reading. Neither, however, will satisfy the military reader with respect to Saxe's military operations. The best military evaluation of Saxe is contained in Liddell Hart's *Great Captains Unveiled,* in the chapter: *Marechal de Saxe—Military Prophet.* The only comprehensive studies of Saxe's campaigns are: *Les Campagnes du Marechal de Saxe,* by Captain J. Colin, Paris, 1901, and *Maurice de Saxe, Marechal de France,* by General Camon, Paris, 1934.

PREFACE

THIS work was not born from a desire to establish a new method of the art of war; I composed it to amuse and instruct myself.

War is a science replete with shadows in whose obscurity one cannot move with an assured step. Routine and prejudice, the natural result of ignorance, are its foundation and support.

All sciences have principles and rules. War has none. The great captains who have written of it give us none. Extreme cleverness is required even to understand them. And it is impossible to base any judgment on the relations of the historians, for they only speak of war as their imaginations paint it. As for the great captains who have written of it, they have attempted rather to be interesting than instructive, since the mechanics of war is dry and tedious. Books dealing with war have small success and their merit will not be recognized except after the passage of time. Those writing historically of war have better luck; they are sought by all the curious and kept in all libraries. That is why we have only a confused idea of the discipline of the Greeks and Romans.

Ignorance of Principles

Gustavus Adolphus [King of Sweden, 1594-1632] created a method that was followed by his disciples, all of whom accomplished great things. But since his time there has been a gradual decline amongst us, which must be imputed to our having learned only his forms, without regard to principles. Hence the confusion of customs, these having been added to or detracted from according to fancy. But in reading Montecuculli [famous Austrian general, of Italian birth, 1609-1680], who was contemporary with Gustavus and is the only general who entered into some detail, it is very evident that we have departed already more from his methods than he did from those of the Romans. Thus there remain nothing but customs, the principles of which are unknown to us.

Chevalier Follard has been the only one who has dared to pass the bounds of these prejudices; I approve his noble courage. Nothing is so disgraceful as slavishness to custom; this is both a result of ignorance and a proof of it. But Chevalier Follard goes too far. He advances an opinion which he pronounces infallible, without reflecting that success depends upon an infinite number of circumstances which human prudence cannot foresee. He supposes all men to be brave at all times and does not realize that the courage of the troops must be reborn daily, that nothing is so variable, and that the true skill of a general consists in knowing how to guarantee it by his dispositions, his positions, and those traits of genius that characterize great captains. Perhaps he reserved discussion of this immense subject; and perhaps, also, it escaped him. Nevertheless, it is of all the elements of war the one that is most necessary to study.

The same troops, who if attacking would have been victorious, may be invariably defeated in entrenchments. Few men have accounted for this in a reasonable manner, for the solution lies in human hearts and one should search for it there. No one has written of this matter which is the most important, the most learned and the most profound, of the profession of war. And without a knowledge of the human heart, one is dependent upon the favor of fortune, which sometimes is very inconstant. I shall make use of one example to reinforce my opinion.

Defeat After Victory

After the French infantry had repulsed the Imperialists at the battle of Friedelingen [1702, War of the Spanish Succession, 1701-14] with incomparable valor, after they had routed them several times and had pursued them through a wood and onto a plain which lay on the other side, someone cried that they were cut off—two troops had appeared (and these may have been French). All this victorious infantry fled in frightful disorganization, although no one either attacked or pursued them, repassed through the wood, and only halted on the other side of the battlefield. Marshal de Villars [1653-1734] and the generals tried to rally them in vain. However, the battle had been won; the French cavalry had defeated that of the Imperialists, and no more of the enemy were to be seen.

Nevertheless, it was those same men who at one moment had defeated the Imperial infantry with the utmost intrepidity and at another had been seized with such a panic of terror that it was almost impossible to regain their courage. Marshal de Villars himself told this to me when he was showing me the plans of his battles at Vaux-Villars. Anyone who wishes to search for similar examples will find numerous in the history of all nations. This one, however, is sufficient to prove the instability of the human heart and how little we should depend on it. But before enlarging too much upon the higher parts of war, it will be necessary to treat of the lesser, by which I mean the foundations of the art.

Detailed Knowledge Vital

Although those who occupy themselves with details are considered to be men of limited capacity, it seems to me, nevertheless, that this part is essential, because it is the foundation of the profession, and because it is impossible to erect any edifice, or to establish any system, without first knowing the principles that suport it. I shall illustrate my meaning with a comparison. A man who has a talent for architecture and can design, will draw the plan and perspective of a palace with great skill. But if he is to execute it, if he does not know how to shape his stones, to lay his foundation, the whole edifice will soon crash. It is the same with the general who does not know the principles of his art, nor how to organize his troops, for these are indispensable qualifications in all the operations of war.

The prodigious success which the Romans always gained with small armies against multitudes of barbarians can be attributed to nothing but the excellent composition of their troops. Not that I would infer from this that a man of genius will not be able to succeed, even at the head of an army of Tartars. It is much easier to take men as they are than to make them as they should be; it is difficult to reconcile opinions, prejudices, and passions.

I shall commence with our system of raising troops; then I shall examine how to supply, train, and fight them.

It would be foolhardy to state that all the methods now employed are worthless, for it is a sacrilege to attack usages, albeit one less great than to propose something new. Therefore, I declare that I shall only attempt to indicate the abuses into which we have fallen.

I

RAISING TROOPS

TROOPS are raised by enlistment with a fixed term, without a fixed term, by compulsion sometimes, and most frequently by tricky devices.

When recruits are raised by enlistment it is unjust and inhuman not to observe the engagement. These men were free when they contracted the enlistment which binds them, and it is against all laws, human or divine, not to keep the promises made to them. What happens when promises are broken? The men desert. Can one, with justice, proceed against them? The good faith upon which the conditions of enlistment were founded has been violated. Unless severe measures are taken, discipline is lost; and, if severe punishments are used, one commits odious and cruel acts. There are, however, many soldiers whose term of service is ended at the commencement of a campaign. The captains who wish to have their organizations complete retain them by force. This results in the grievance of which I am speaking.

The raising of troops by fraud is also an odious practice. Money is slipped secretly into the man's pocket and then he is told he is a soldier.

Raising troops by force is still worse. This is a public misfortune from which citizens and inhabitants of a country can only save themselves by bribery, and it is founded on shameful methods.

Argument for Conscription

Would it not be better to prescribe by law that every man, whatever his condition in life, should be obliged to serve his prince and his country for five years? This law could not be objected to, because it is natural and just that all citizens should participate in the defense of the nation. No hardship could result if they were chosen between the ages of twenty and thirty years. These are the years of relative freedom of action, so far as the individual is concerned, when youth seeks its fortune, travels, and is of little comfort to parents. This would not be a public calamity because one could be sure that, when five years had passed, discharge would be granted.

This method of raising troops would provide an inexhaustible reservoir of fine recruits who would not be so much inclined to desert. In course of time, as a consequence, it would be regarded as an honor to have completed one's military service. But to produce this effect it is essential to make no distinctions, to be immovable on this point and to enforce the law particularly on the nobles and the rich. Then, no one will complain. Consequently, those who have served their time will scorn those who are reluctant to obey the law, and insensibly it will become an honor to serve. The poor bourgeois will be consoled by the example of the rich, and the rich will not dare complain upon seeing the noble serve.

Arms is an honorable profession. How many princes have borne arms! Witness M. de Turenne [noted French general, 1611-1675]. And how many officers have I seen serve in the ranks rather than live in indolence! Thus only effeminacy will make this law appear hard to some. But everything has a good and a bad side.

II

CLOTHING TROOPS

OUR uniform is not only expensive, but very uncomfortable; the soldier is neither well shod, clothed, nor quartered. The love of appearance prevails over attention to health, and this is one of the most important points demanding our attention. Long hair is an unseemly appendage of a soldier; and once the rainy season has arrived, his head is seldom dry.

As for his feet, it is not to be doubted that his stockings, shoes, and feet rot together, since he has no extra pairs to change; and even if he should have, they will be of little use to him because quickly he will be back in the same state. Thus the poor soldier is soon sent to the hospital. White leggins spoil in washing, are good only for reviews, are inconvenient and harmful, of no real use, and very expensive.

The hat soon loses its shape and cannot resist the mistreatment of a campaign. Soon it no longer sheds rain, and as soon as the soldier lies down it falls off. The soldier, worn out with fatigue, sleeps in the rain and dew with his head uncovered and the next day has a fever.

Helmets Better Than Hats

In place of hats, I should prefer helmets. They do not weigh more than hats, are not at all uncomfortable, protect from a saber blow, and are sufficiently ornamental.

I should like to have the soldier clothed in a jacket, a little large, with a small one under it, something like a vest, and a Turkish coat with an attached hood. These coats cover a man well and do not contain more than two ells and one half of cloth. They are light and cheap.

The soldier will have his head and neck covered from the rain and wind. Lying down, he will be relatively dry because these clothes are not tight. When wet they will dry quickly in fair weather.

It is quite different with the present uniform. As soon as it is wet, the soldier feels it to the skin, and it must be dried on him. One need not, therefore, be astonished to see so much disease in the army. The strongest resist the longest, but in the end they usually succumb. If one adds to what I have said the service that those who are well are obliged to do in place of the sick, the dead, the wounded, and the deserters, it is not astonishing to see battalions reduced to a hundred men at the end of a campaign. That is how the smallest things influence the greatest.

Shoes Instead of Boots

As for shoes, I would prefer the soldiers to have shoes of thin leather with low heels, instead of heavy boots. They would be perfectly shod and would march with better grace, since the low heels would force them to turn their toes out, stretch their joints, and consequently, draw in their shoulders. The shoes should be worn on the bare feet and the feet greased with tallow or fat. This may sound strange, but the French veterans did this, for experience proved that they never blistered their feet, and water did not soak the shoes easily on account of the grease. On the other hand the leather did not get hard and hurt the feet.

The Germans, who make their infantry wear woolen stockings, have always had numbers crippled by blisters, ulcers, and all sorts of diseases of the feet and legs, because wool is poisonous to the skin. Besides, these stockings soon break through the toes, remain wet, and rot with the feet.

To keep the feet dry, wooden sandals should be added to the footwear. This will keep the shoes from getting wet in the mud and dew and when on duty—a great nuisance and resulting in illnesses. In dry weather, for combat and for drill, they would not be used.

III

FEEDING TROOPS

AS I WOULD divide my troops into centuries, a sutler should be detailed for each century. He should have four wagons drawn by two oxen each and should be provided with a large kettle to hold soup for the whole century. Every man would receive his portion in a wooden dish, together with boiled meat at noon and roast meat at night. It should be the officers' duty to see that they are not imposed on and have no occasion for complaint.

The profit allowed the sutlers would come from the sale of liquor, cheese, tobacco, and the skins from the cattle they killed. The sutlers should provide food for the cattle, and when the army is near forage they would be given the necessary orders for obtaining it.

It might appear difficult to arrange this at first. But with a little care everyone would be satisfied. When soldiers are detached in small parties they could carry two days' supply of roast meat with them, without encumbrance. The quantity of wood, water, and kettles to make soup for a hundred men is more than would be sufficient for a thousand the way I propose, and the soup made in the customary manner is never as good. Besides, the soldiers eat all sorts of unhealthy things, such as pork and unripe fruit, which make them ill. The officer would only have to watch a single kitchen, that of the sutler, and an officer should always be present at each meal to see that the soldiers had no cause for complaint.

On forced marches, when the baggage could not be brought up, cattle would be distributed to the troops. Wooden spits could be made and the meat roasted. This would not be inconvenient and would only last a few days. Let my method and the former one be balanced. I am persuaded that mine will be found the better.

The Turks do this and they are perfectly well nourished. One can tell their cadavers, after battles, from the German's, which are pale and emaciated. This has at times another advantage; the soldier's money may be saved by giving them their entire pay and selling their food to them. There are certain countries, like Poland and Germany, where cattle abound. Contributions are demanded from the inhabitants and, to enable them to sustain them, they are taken half in food and half in money. The food is sold to the soldiers; thus their pay is in continual circulation, and one has money as well as contributions, an important consideration.

Biscuits Better Than Bread

Soldiers should never be given bread in the field, but should be accustomed to biscuit, because it will keep for fifty years or more in depots and a soldier can easily carry a fifteen days' supply of it. It is healthy; one needs only inquire of the officers who served with the Venetians to learn the advantages of biscuit. The Russian biscuit, called *soukari,* is the best of all because it does not crumble; it is square and of convenient size. Fewer wagons are needed to carry it than bread.

Soldiers at times should be accustomed to do without biscuit and should be given grain and taught to cook it on iron plates, after having ground and made it into paste with water. Marshal Turenne said something about this in his *Memoirs.*

I have heard of great captains who, even when they had bread, did not allow the troops to eat it, so as to accustom them to do without it. I have made eighteen-month campaigns with troops who were habituated to do without bread and without hearing complaints. I have made several others with troops who were accustomed to it, and they could not do without it. Let bread be lacking a single day and there was trouble. The result was that not a step ahead could be taken, nor any bold march.

Vinegar for Health

I should not omit to mention here a custom of the Romans by which they prevented the diseases that attack armies with changes of climate. A part of their amazing success can be attributed to it. More than a third of the German armies perished upon arrival in Italy and in Hungary. In the year 1718 we entered the camp at Belgrade with 55,000 men. It is on a height, the air is healthy, the spring water good, and we had plenty of everything. On the day of battle, August 18, there were only 22,000 men under arms; all the rest were dead or unable to fight.

I could produce similar instances among other nations. It is the change of climate that causes it. There were no such examples among the Romans as long as they had vinegar. But just as soon as it was lacking they were subject to the same misfortunes that our troops are at present. This is a fact to which few persons have given any attention, but which, however, is of great consequence for the conquerors and their success. As for how to use it, the

Romans distributed several days' supply among their men by order, and each man poured a few drops in his drinking water. I leave to the doctors the discovery of the causes of such beneficial effects; what I report is unquestionable.

IV

PAY OF TROOPS

WITHOUT going into detail about the different rates of pay, I shall say only that it should be ample. It is better to have a small number of well-kept and well-disciplined troops than to have a great number who are neglected in these matters. It is not big armies that win battles; it is the good ones. Economy can be pushed only to a certain point. It has limits beyond which it degenerates into parsimony. If your pay and allowances for officers will not support them decently, then you will only have rich men who serve for pleasure or adventure or indigent wretches devoid of spirit.

For the first I have little use because they can stand neither discomfort nor the rigor of discipline. Their talk is always seditious, and they are nothing but frank libertines. As for the others, they are so depressed that one can expect no great virtue in them. Their ambition is limited because advancement hardly interests them; and worse, they prefer to retain whatever rank they have, especially when promotion is an expense.

Inspiration in Hope and Poverty

Hope encourages men to endure and attempt everything; in depriving them of it, or in making it too distant, you deprive them of their very soul. It is essential that the captain should be better paid than the lieutenant, and so for all grades. The poor gentleman should have the moral surety of being able to succeed by his acts and his services. When all these things are taken care of you can maintain the most austere discipline among the troops.

Truly, the only good officers are the poor gentlemen who have nothing but their sword and their cape, but it is essential that they should be able to live on their pay. The man who devotes himself to war should regard it as a religious order into which he enters. He should have nothing, know no other concern than his troop and should hold himself honored in his profession.

In France, a young noble considers himself badly treated by the court if a regiment is not confided to him at the age of eighteen or twenty years. This practice destroys all spirit of emulation in the rest of the officers and in the poor nobility, who are almost certain never to command a regiment, and, in consequence, to gain the more important posts wherein glory is a recompense for the trouble and suffering of a laborious life to which they are willing to sacrifice themselves if confident of a flattering and distinguished career.

I do not argue by this that no preference should be shown to some princes or others of illustrious rank, but it is essential that these marks of preference should be justified by distinguished merit.

V

DRILL

DRILL is necessary to make the soldier steady and skillful, although it does not warrant exclusive attention. Among all the elements of war it is even the one that deserves the best, if one excepts those which are dangerous.

The foundation of training depends on the legs and not the arms. All the mystery of maneuvers and combats is in the legs, and it is to the legs that we should apply ourselves. Whoever claims otherwise is but a fool and possessor only of elementary knowledge of what is called the profession of arms.

The question whether war is a trade or a science is

defined very well by Chevalier Follard. He said: "War is a trade for the ignorant and a science for the expert."

VI

FORMING TROOPS FOR COMBAT

THIS is a broad subject and I propose to deal with it in a manner so different from respected custom that I shall probably expose myself to ridicule. But to lessen the danger, I shall explain the present method. This is no small affair, for I could compose a big volume on it.

I shall begin with the march, and this makes it necessary to say something that will appear highly extravagant to the ignorant. No one knows what the ancients meant by the word *tactics*. Nevertheless, many military men use this word constantly and believe that it is drill or the formation of troops for battle. Everyone has the march played without knowing how to use it. And everyone believes that the noise is a military ornament.

We should have a better opinion of the ancients and of the Romans, who are our masters in warring and who deserve to be. It is absurd to imagine that their martial music, trumpet calls and the like had no purpose other than to confuse the soldiers.

But to return to the march, about which everyone bothers themselves to distraction, but will never reach a conclusion unless I reveal the secret. Some wish to march slowly, others would march fast. But what about the troops whom no one knows how to make march fast or slow, as they desire, or as becomes necessary, and who require an officer at every corner to make them turn, some like snails and others running, or to advance a column which is always trailing?

It is a comedy to see even a battalion commence movement! It is like a poorly constructed machine, about to fall apart at every moment and which straggles on with infinite difficulty. Do you wish to hurry the head? Before the tail knows that the head is marching fast, intervals have been formed. To close up the column, the tail must run; the head that follows this tail must do the same. Soon everything is in disorder, with the result that you are never able to march rapidly.

Marching to Music

The way to remedy all these inconveniences, and many others which follow and are of greater importance, is very simple, nevertheless, because it is dictated by nature. Shall I say it, this great formula which comprises all the mystery of the art and which will no doubt seem ridiculous? Have them march in cadence. There is the whole secret, and it is the military step of the Romans. That is why these musical marches were instituted, and that is why one beats the drum; it is this which no one knows and which no one has perceived.

With this you can march fast or slow as you wish; the tail will not lose distance; all your soldiers will start on the same foot; the changes of direction will be made together with speed and grace; the legs of your soldiers will not get tangled up; you will not be forced to halt after each turn in order to start off on the same foot; your soldiers will not exhaust themselves a quarter as much as at present. All this may seem extraordinary. Every one has seen people dancing all night.

But take a man and make him dance for a quarter of an hour only without music, and see if he can bear it. This proves that music has a secret power over us, that it predisposes our muscles to physical exercise and lightens the exercise.

If anyone asks me what music should be played for men to march by, I should answer, and seriously, that all marches in double or triple time are suitable for it, some more, others less, depending upon whether they are more or less accented; that all the airs played on the tamborine or fife will do, likewise, and that one needs only to choose the more suitable.

I shall be told, perhaps, that many men have no ear for music. This is false; movement to music is natural and is automatic. I have often noticed, while the drums were beating for the colors, that all the soldiers marched in cadence without intention and without realizing it. Nature and instinct did it for them. I shall go further: it is impossible to perform any evolution in close order without music or drums, and I shall prove it in the proper place.

Romans Marched Rapidly

Considering what I have said superficially, it does not appear that this cadence is of such great importance. But, in a battle to be able to augment the rapidity of march, or to diminish it, has infinite consequences. The military step of the Romans was nothing else; with it they marched twenty-four miles in five hours, the equivalent of eight leagues. Let anyone try the experiment on a body of our infantry and see if it is possible to make them do eight leagues in five hours. Among the Romans this was the principal part of their drill. From this, one can judge the attention they gave to keeping their troops in condition, as well as the importance of cadence.

What will be said if I prove that it is impossible to charge the enemy vigorously without this cadence, and that without it one reaches the enemy always with ranks opened? What a monstrous defect! I believe, however, that no one has given it any attention for the past three or four hundred years.

It now becomes necessary to examine a little our method of forming battalions and of fighting. The battalions are in contact with each other, since the infantry is all together and the cavalry also (for which, in truth, there is no common sense—but this will be covered in the proper place). The battalions, then, march ahead, and this very slowly because they are unable to do anything else. The majors cry: "Close," on which they press toward the center; insensibly the center gives way, which makes intervals between the battalions. There is no one who has had anything to do with these affairs but will agree with me.

The majors' heads are turned because the general, whose head is turned also, cries at them when he sees the space between his battalions and is fearful of being taken in the flanks. He is thus obliged to call a halt, which should cost him the battle; but since the enemy is as badly disposed as he, the harm is slight.

Evil of Bad Formations

A man of intelligence would not stop to repair this confusion, but would march straight ahead, for if the enemy moves he would be lost. What happens? Firing commences here and there, which is the height of misfortune. Finally the opposing forces approach each other, and one of the two ordinarily takes to their heels at fifty or sixty paces, more or less. There you have what is called a charge. What is the cause of it all? Bad formations make it impossible to do better.

But I am going to suppose something impossible; I mean two battalions attacking to march toward each other without wavering, without doubling, and without breaking. Which will gain the advantage? The one that amuses itself shooting, or the one which will not have fired? Skillful soldiers tell me that it will be the one that has held its fire, and they are right. For besides being upset when he sees his opponent coming at him through the smoke, the one who has fired must halt to reload. And the one who stops while the other is marching toward him is lost.

If the previous war had lasted a little longer, indubitably everyone would have fought hand to hand. This was because the abuse of firing began to be appreciated; it causes more noise than damage, and those who depend on it are always beaten.

Ineffective Musket Fire

Powder is not as terrible as is believed. Few men, in these affairs, are killed from in front or while fighting. I have seen entire salvos fail to kill four men. And I have never seen, and neither has anyone else, I believe, a single discharge do enough violence to keep the troops from continuing forward and avenging themselves with bayonet and shot at close quarters. It is then that men are killed, and it is the victorious who do the killing.

At the battle of Calcinato, M. de Reventlaü, who commanded the Imperial army, had ranged his infantry on a plateau and had ordered them to allow the French infantry to approach to twenty paces, hoping to destroy them with a general discharge. His troops executed the orders exactly.

The French with some difficulty climbed the hill which separated them from the Imperials and ranged themselves on the plateau opposite the enemy. They had been ordered not to fire at all. And since M. Vendome did not care to attack until he had taken a farm which was on his right, the troops remained for a considerable time looking at each other at close range. Finally they received the order to attack.

The Imperials allowed them to approach to within twenty or twenty-five paces, raised their arms, and fired with entire coolness and with all possible care. They were broken before the smoke had cleared. There were a great many killed by point blank fire and bayonet thrusts and the disorder was general.

Turks' Winning Tactics

At the battle of Belgrade [August 16, 1717], I saw two battalions cut to pieces in an instant. This is how it happened. A battalion of Neuperg's and another of Lorraine's were on a hill that we called the battery. At the moment when a gust of wind dissipated the fog which kept us from distinguishing anything, I saw these troops on the crest of the hill, separated from the rest of our army.

Prince Eugene, [of Savoy, French general in Austrian service, 1663-1736; fought with Marlborough, 1704-9], at the same time discovering a detachment of cavalry in motion on the side of the hill, asked me if I could distinguish what they were. I answered that they were thirty or forty Turks. He said: "Those men are enveloped," speaking of the two battalions. However, I could perceive no sign of their being attacked, not being able to see what was on the other side of the hill. I hastened there at a gallop.

The instant I arrived behind Neuperg's colors, the two battalions raised their arms and fired a general discharge at thirty paces against the main body of the attacking Turks. The fire and the meleé were simultaneous, and the two battalions did not have time to flee for every man was cut to pieces on the spot. The only persons who escaped were M. Neuperg, who, fortunately for him, was on a horse; an ensign with his flag who clung to my horse's mane and bothered me not a little, and two or three soldiers.

At this moment Prince Eugene came up, almost alone, being attended only by his body guard, and the Turks retired for reasons unknown to me. It was here that Prince Eugene received a shot through the sleeve. Some cavalry and infantry arriving, M. Neuperg requested a detachment to collect the clothing. Sentries were posted at the four corners of the ground occupied by the dead of the two battalions, and their clothes, hats, shoes, etc., were collected in heaps. During this ceremony, I had curiosity enough to count the dead; I found only thirty-two Turks killed by the general discharge of the two battalions—which has not increased my regard for infantry fire.

Defects of Large Battalions

It was an established maxim with the late M. de Greder, a man of reputation, and who had for a long time commanded my regiment of infantry in France, to make his men carry their muskets shouldered in an engagement; and, in order to be still more master of their fire, he did not even permit them to make their matches ready. Thus he marched toward the enemy, and the moment they started to fire, he threw himself sword in hand at the head of the colors, and cried, "Follow me!" This always succeeded for him. It was thus that he defeated the Frisian Guards at the battle of Fleurus [July 1, 1690].

What I have been advancing appears to me supported by reason and experience and proves that these large battalions have terrible defects; they are good only to fire, and they are organized for that alone. When fire is useless, they are worth nothing and have no recourse but to save themselves; which proves that everything from its very nature falls to its own level.

Shall I tell from where I think we gained this method? It was probably taken from parades. This manner of arrangement makes a more pleasing appearance; unwittingly we have become accustomed to it, so that it was adopted in action.

Some attempt to vindicate this ignorance, or forgetfulness of good things, by apparent reason, alleging that in thus extending their front they will be able better to employ their fire. I have even known some to draw up their battalions three deep, but misfortune has been the fate of those who have done it. Otherwise, I really believe (God forgive me) they soon would have formed them two deep, and not improbably in single rank. For all my life I have heard it said that one should extend his order to out-flank the enemy. What absurdity!

But enough of this. I must first describe my method of forming regiments, legions and the cavalry, because it is essential to base oneself on principle and on a formation for combat, which can change with the variety of situations but which will not be destroyed.

VII

FORMATION OF THE LEGION

THE Romans vanquished all nations by their discipline; they meditated on war continually, and they always renounced old customs whenever they found better. In this respect, they differed from the Gauls, whom they defeated during several centuries without the latter thinking of correcting their errors.

Their legion was a body so formidable as to be capable of undertaking the most difficult enterprizes. "It was undoubtedly a god," says Vegetius,* "that inspired the legion." I have held the same opinion for a long time, and it is this which has rendered me more aware of the defects of our own practices.

Since what I write is only a diversion to dissipate my boredom, I want to give full play to my imagination.

I would form my body of infantry into legions, each composed of four regiments, and every regiment of four centuries; each century would have a half-century of light-armed foot and a half-century of cavalry.

When centuries of infantry are drawn up in separate bodies, I shall call them battalions, and the cavalry squadrons, in order to conform to our usage and aid the interchange of ideas.

The centuries, both of foot and horse, are to be composed of ten companies, every company consisting of fifteen men.

It is necessary in a monarchy to adapt the state of the troops to economy. It therefore is expedient to form them in three different establishments, which I shall call: the establishment in peace, the establishment for war, and the full establishment for war.

When the country is completely at peace the companies

**The Military Institutions of the Romans*, by Vegetius, The Military Service Publishing Company, Harrisburg, Pennsylvania.

are to consist of one sergeant, one corporal, and five veteran soldiers [a cadre]. When preparations are being made for a war that is expected, an addition of five men is to be made. When war has been declared, or is about to be declared, they should consist of one sergeant, one corporal and fifteen men, which is an increase of 1600 per legion. The five veterans per company will constitute a reserve for the occasional supply of officers and noncommissioned officers, for these are always difficult to develop. In addition, among the five veterans there will always be a reserve for replacement. I do not care for newly raised regiments; sometimes they are worth nothing after ten years of war.

Veteran Men and Horses

As for the cavalry, it should never be touched; the old troopers and the old horses are good, and recruits of either are absolutely useless. It is a burden, it is an expense, but it is indispensable.

In regard to the infantry, as long as there are a few veterans you can do what you want with the rest; they are the greatest number, and the return of these men in peace is a noticeable benefit to the nation, without serious diminution to the military forces.

As I am going to deal with war, I shall place my troops on a complete war establishment, so that a century of foot will consist of 184 men, and each company of seventeen.

The two half-centuries of horse and light-armed foot should not exceed ten per company, including the sergeants and corporals, because they will be recruited out of the regiments.

Any diminution of the heavy-armed forces, which compose the main body of the infantry, will be of no consequence, because the different units of the legion will still remain the same even if reduced to peace strength, that is to say, five soldiers. This will be a great advantage and a solid foundation for all your infantry, since your drills remain the same and permanent.

"CENTURY" HAS 10 COMPANIES EACH OF 15 MEN. Σ (15×10) = 150 MEN
16 CENTURIES = 16×150 = 2400
REGTS = 2400/4 = 600 MEN
REGT WILL HAVE 300 INF (BN), 300 CAV (SQDN)

It is inconceivable how prejudicial all changes are. I have seen troops belonging to the same government, when assembled after a long peace, differ to such a degree in their maneuvers and formation of their regiments that one would have taken them for a collection from several distinct nations.

It is necessary, therefor, to establish one definite principle of action, and never to depart from it. No one should be ignorant of this principle because it is the foundation of the military profession. But it is impossible to assure it, unless you always maintain the same number of officers and noncommissioned officers; without that, your maneuvers will always vary.

VIII

ARTILLERY, SMALL ARMS

EVERY heavy-armed century is to be furnished with an arm of my own invention, which I call an *amusette.* They carry more than four thousand paces with extreme velocity. [12,000 feet, or more than two miles, probably an exaggerated estimate]. The field-pieces used by the Germans and Swedes with their battalions will scarcely carry a fourth of that distance.

This is also much more accurate; two men can carry it anywhere. It fires a half-pound lead ball and one hundred pounds of projectiles are carried with it. Going through foot paths in mountains, the trails are drawn back and two soldiers can carry the piece very easily. This arm can be used on a thousand occasions in war.

The artillery and wagons should be drawn by oxen. The wagon should be loaded with all kinds of implements necessary for building forts, such as different cordages, cranes, pullies, windlasses, saws, hatchets, shovels, mattocks, etc. These should all be marked with the number of their respective legion, so that in armies they will not be lost or mixed together.

The private soldier should have a piece of copper

Hussar, Charging

fixed on each shoulder, with the number of the legion and regiment to which they belong, respectively, on them so that they may be easily distinguished.

Identification Marks

I would also have their right hands marked with the same numbers by a composition used by the Indians, which never wears off and will put a stop to desertion. This will be easy to introduce and will lead to innumerable good consequences. To establish it the sovereign has only to assemble his colonels and tell them that it will be of great importance in maintaining good order and preventing desertion; that they will please him if they give the example and mark themselves; that this could not be other than a mark of honor, by proving the regiment in which they have served. No one will refuse it. All the subordinate officers, ambitious to oblige their Prince, and realizing the utility of such an institution, will gladly imitate their colonels. After this no soldier will refuse it. Marking them could even be made a ceremony. It was a practice among the Romans, but they marked with a hot iron.

For the centuries of horse, men should be chosen from the regiments to which they belong, leaving the choice of them to their centurion, but he will give preference to the old soldiers. Cavalry thus selected will never abandon their infantry and will give them confidence in a battle and be of admirable service to them, either in pursuit or covering their retreat. But I shall speak more fully of this in another place.

Breechloaders Favored

The light-armed foot are in like manner to be chosen in their regiments, the centurions selecting the youngest and most active. Their arms must consist of nothing more than a very light fowling-piece and bayonet with a handle to it. This fowling-piece is to be made so as to open and receive the charge at the breech, so that it will not need to be rammed [the breech loading piece was not successfully developed until long after Saxe's time]. All the equipment must be as light as possible. Their officers will be chosen in the same manner without regard to seniority. They must be drilled frequently, must practice jumping and running, but, above all, firing at a mark at three hundred paces distance. Rewards are to be posted for those who excel in all these different exercises, in order to create emulation.

A body of infantry organized according to this plan, and thoroughly trained, can march everywhere with the cavalry and, I am confident, will be capable of giving great service.

I am far from approving of grenadiers; they are the elite of our troops. Since they are employed on every important occasion, a brisk war exhausts them to such an extent that they are no longer able to furnish noncommissioned officers, who are the heart of the infantry. I would substitute veterans in place of grenadiers; they should have more pay than the simple soldiers and light foot. The light-armed forces are to be employed on all services requiring speed and activity, the veterans only on serious efforts. I believe this will result in great benefit for the military establishment.

Grenadiers In Demand

The command of the light-armed troops is always to be given to a lieutenant, who should be chosen by the colonel. But that of the veterans, being regarded as the post of honor, is to be determined by seniority. According to the present system, it is impossible to prevent the officers from succeeding to grenadier companies by seniority without seriously affronting them. This always uses up the best officers you have. I have seen sieges where the companies of grenadiers had to be replaced several times. This is easily explained: grenadiers are wanted everywhere. If there are four cats to chase, it is grenadiers who are demanded, and usually they are killed without any necessity.

Infantryman, Regiment of the Dauphin

The heavy-armed forces are to have good muskets, five feet in length, with large bores and using a one ounce ball. These muskets also should load at the breech. They will carry over twelve hundred paces.

It is needless to be afraid of over-loading the infantry with arms; this will make them more steady. The arms of the Roman soldiers weighed over sixty pounds, and it was death to abandon them in action. It prevented any thought of flying and was a principle of military art with them. To these muskets I would add a bayonet with a handle, two and one-half feet long, which will serve as a sword, and oval shields or targets. These shields have many advantages; they not only cover the arms, but, when fighting in position, the troops can form a kind of parapet with them in an instant by passing them from hand to hand to the front. Two of them, one on the other, are musket proof. My opinion in regard to this piece of armor is supported by that of Montecuculli, who says that it is absolutely necessary for the infantry.

Bayonets and Swords

Bayonets with handles to fix within the barrel of the muskets are much preferable to the others because they enable the commander to reserve his fire as long as he thinks proper. This is a matter of the utmost importance, since one cannot hope to do two different things at once. That is to say, charge, or stand and fight. In one case they must fire, in the other not at all.

Here is an example: Charles the XII, King of Sweden, wished to introduce among his troops the practice of engaging sword in hand. He had spoken of it several times, and the army knew that he favored the system. Accordingly, in a battle against the Russians, at the moment it was about to begin he hastened to his regiment of infantry, and made a spirited harangue, dismounted, posted himself in the front of the colors, and led them on to the charge himself. But as soon as they came within about thirty paces of the enemy, his whole regiment fired

Musketeer, Firing Position

in spite of his orders and his presence. And although he routed the Russians and obtained a complete victory, he was so piqued that he passed through the ranks, remounted his horse, and rode off without speaking a single word.

IX

INFANTRY FORMATION

BUT to return to the formation of the battalions. I would draw them up four deep, the two front ranks being armed with muskets and the two rear with half-pikes and muskets slung over their shoulders. The half-pike is a weapon thirteen feet in length, exclusive of the iron-head, which is to be trangular, eighteen inches long and two broad. The shaft must be of spruce, hollowed, and covered with varnished parchment. This is very strong and very light and does not whip like the pikes the infantry think they cannot do without.

I have always heard this opinion from all experienced men; and the same reasons, neglect and indolence, which have caused the abandonment of many other excellent customs of the profession of war, are the cause of the abandonment of this one. The half-pikes were found unserviceable in some affairs that took place in Italy, where the country was very rough; since then they have been laid aside everywhere and nothing since has been thought of but to increase the quantity of firearms and to fire.

Although I have been exclaiming against firing in general, there are certain situations where it is necessary and it is well to know how. These are enclosures and rough grounds, and also against cavalry. But the method should be simple and natural. The present practice is worthless because it is impossible for the soldier to aim while his attention is distracted awaiting the command. How can all these soldiers who have been commanded to get ready to fire continue to aim until they receive the word to fire? A trifle will derange them, and, having

Musketeer, Fixing Bayonet

once lost the critical instant, their fire is no longer of much use. Let no one think that this does not make a great difference; it will amount to several yards. Nothing is so easy to derange as musket fire. And besides this, and according to our method, they are kept in a constrained position.

These and many other inconveniences totally prevent the effect which small-arms should produce. But this subject demands a special article. I shall therefore return to the formation of my battalions.

Pikes and Muskets

In attacking infantry, the two rear ranks are to lower their pikes; in this position the pikes will extend from six to seven feet ahead of the front rank. The front ranks being sheltered in such a manner will, I am sure, aim with more confidence than if they had nothing in front of them. Besides this, the third rank can ward off blows and defend the first rank, which it will do much better, since it is covered by the first two ranks. The second rank, which is armed with muskets, can fire and defend the man in front of him in the first rank, without the latter being obliged to stoop.

This avoids a serious disadvantage which is incurred in kneeling, a dangerous movement, because men who are afraid prefer this position, they cannot be made to get up when wanted, and it is always necessary to halt to kneel. According to my formation all the men are covered, each by the other, with reciprocal confidence; the front presents a forest of spears; their appearance is formidable and gives confidence to your own troops because they feel its power.

In attacking infantry, the light-armed foot are to be dispersed along the front, at the distance of a hundred, one hundred fifty, or two hundred paces in advance. They should begin firing when the enemy is about three hundred paces off, without word of command and at will, until the enemy approaches within about fifty paces.

At this distance every captain is to order a retreat, taking care to retire slowly towards his regiment, keeping up his fire from time to time, until he arrives at his battalion which should be starting to move. The men should be disposed to fall into the intervals of the battalions by tens. The regiments during this time should have doubled ranks while moving forward. There should be two troops of cavalry, of thirty troopers each, thirty paces behind the regiment.

Discomforting the Enemy

The whole moving forward with a regular and rapid step will certainly discourage the enemy. For what can they do? To attack the flanks of the centuries they must break their battalions. They neither can nor dare because the intervals are only ten paces and these are filled by the light-armed foot. In addition they are rendered still more impenetrable by the transversed pikes of the rear ranks. How can they resist, being only four deep, and having been already harassed by the light-armed infantry, when they meet fresh troops formed eight deep, with a front equal to theirs, and which come rapidly against them, disordered already by the unevenness of ranks which is unavoidable in the movement of so extensive a body? It appears highly probable that they must be defeated, and if they trust to flight they are lost beyond recovery.

For the moment they turn their backs, the light-armed foot, together with the horse posted in the rear, are to pursue, and will make dreadful havoc among them. The seventy cavalry and the seventy light-armed foot should destroy a battalion in a moment before they have had time to flee a hundred paces. During the pursuit the centuries are to stand fast, in order to receive their own troops again and to be ready to renew the charge.

I cannot avoid believing that, of all formations, this is the best for battle. Some will say that the enemy's cavalry might be thrown against my light-armed troops.

Basque Volunteer

No one will dare it, but so much the better if it happens. Will they not be forced to withdraw? Can they fire against seventy men scattered along the front of my regiment? It would be like firing at a handful of fleas. Ah! They will do the same thing and also will have light-armed foot. The benefit of my system is proved if it bothers them to the point where they are forced to imitate it.

Fire of Light-armed Troops

I should, before I finish this article, make a concise calculation of the fire of my light-armed troops.

Let us suppose them to begin firing at the distance of three hundred paces, which is that at which they are trained, and that they are one hundred fifty paces from me. They thus will fire during the time necessary for the enemy to march that distance, which will be from seven to eight minutes at least. My irregulars will be able to fire six times in a minute. However, I shall only say four; every one will, therefore, have fired thirty times. Consequently, every battalion will have received four or five hundred shots before the engagement can possibly commence.

And from whom? From troops who have spent their life firing at a greater distance, who are not drawn up in close order, and who fire at ease without waiting for the word of command. They are not kept in that constrained attitude which is customary in the ranks, where the men crowd one another, and prevent their taking a steady aim. I contend that a single shot from one of these irregulars is worth ten from any other. And if the enemy marches in line, they will receive ten thousand musket shots per battalion before my troops attack them.

To these I add the fire of my *amusettes.* I have already observed that they require only two soldiers to draw and one to serve them.

Before an engagement these *amusettes* are to be advanced in front, along with the light-armed troops. Since they can be fired two hundred times in an hour with ease, and carry above three thousand paces, they cause great damage to the enemy when forming or after they have passed wood, defile, or village. Even when there are none of these obstacles, they will have to march in column, and then draw up in order of battle, which sometimes takes several hours. Every century is to have only one; those of both lines may be joined upon occasion, or all can be collected on a height. They should produce considerable effect because they will carry farther and are much more accurate than our cannon. Since there are four per regiment there will be sixteen to each legion. The sixteen belonging to a legion assembled in an engagement will be sufficient to silence any battery of the enemy's which bothers the cavalry or even the infantry.

With regard to my pikes, if they become useless in rough or mountainous places the soldiers have nothing more to do than to lay them aside for an hour or two. My soldiers can use their muskets which they always carry slung over their shoulders. To say that carrying them will be too great an incumbrance is an objection to which no reply is necessary. Are they not now obliged to carry their tent poles? Nothing more is required than to substitute these pikes, which are better. Their appearance above the tents will be pleasing and ornamental in a camp. Their weight, including the iron, does not exceed four pounds, because they are hollow. Ordinary pikes weigh about seventeen pounds and are extremely unwieldy.

Rivalry Amongst Legions

I maintain that such a body of troops will be of great service, assuming that the legionary general understands and knows what he should know. If the commander in

Grenadier

chief of an army wants to occupy a post, to obstruct the enemy in their projects or in a hundred different situations which are found in war, he has only to order a particular legion to march. Since it is furnished with everything that can be required to fortify itself, it can soon be reasonably secure from any assault. And in the space of four or five days, it should be ready to sustain a siege and arrest an enemy's army.

This organization of the infantry appears to me the more proper, since it is well proportioned in all its parts. The acquired reputation of any single legion will both make an impression on the others and even on the enemy. Such a body will regard their reputation as a tradition and will always be moved by a desire to surpass that of any other. The exploits of an organization which is denominated by a number are not so soon forgotten as those of one which bears the name of their officers because the officers change and their actions are forgotten with them.

Moreover, it is more the nature of men to be less interested in things which relate to others than about those in which they themselves are concerned. The reputation of an organization becomes personal just as soon as it is an honor to belong to it. This honor is much easier to arouse in an organization that keeps its name than one which carries some one elses, that is, its colonel's who very probably may be disliked.

Many persons do not know why all the regiments which bear the names of provinces in France have always performed so well. They give as the entire reason: "It is *esprit de corps*." This is far from being the real reason, as appears from what I have just been observing. Thus we see how matters of the utmost importance depend on an imperceptible point. Besides, these legions form a kind of military fatherland, where the prejudices of different nations are confounded—an important point for a monarch or for a conqueror. For wherever he finds men, he finds soldiers.

Dragoon

Those who imagine that the Roman legions were composed of Romans from Rome are very much deceived; they came from all the nations in the world. But their composition, their discipline, and their method of fighting were better than those of all other nations. This is why they conquered them all. Neither were they conquered in their turn until this discipline had degenerated among the Romans.

X

CAVALRY IN GENERAL

THE cavalry should be active and mounted on horses inured to fatigue. It should be encumbered with as little baggage as possible, and, above all, should not make that common error of using fat horses. If they could face an enemy every day it would only be the better, for this would soon put them in condition to attempt anything. It is certain that the power of cavalry is not understood. Why? Because of the love we have for fat horses.

I had a regiment of German cavalry in Poland with which I marched more than fifteen hundred leagues in eighteen months. I maintain that this regiment was more fit for service at the end of this time than another supplied with fat horses. But to reach this condition the mounts must be gradually accustomed to hardship and toughened by hunts and violent exercise. This will maintain them in condition and increase their endurance. Likewise, it will make cavalrymen of the troopers and give them a martial bearing. But they must be made to gallop at top speed by squadrons, reaching this point by degrees, and not be drilled gently and at long intervals for fear that the animals may sweat. I insist that unless they are accustomed to hard service they will be more subject to accidents and will never be of much utility.

There should be two kinds of cavalry: light and dragoons. I have no use for any others. Light cavalry and well-equipped dragoons are more useful than hussars. Of the first, which is true cavalry, although much the best, the number must be small because they are very expensive and require special attention. Forty squadrons will suffice for an army of thirty, forty, or fifty thousand men. Their drill should be simple and in masses, free from all effort at too great speed. The essential point is to teach them to fight together and never to disperse. The supplying of main guards is the only duty they should do. Escort, detached, and pursuit duty should never be assigned to them. They should be considered like heavy artillery, which marches with the army. Consequently, they should only serve in combats and battles.

Ideal Troopers and Mounts

For this cavalry, selected men from five feet six to five feet seven inches tall are essential. They should be slender and without fat stomachs. Their horses should be strong and never under fifteen hands, two inches. The German horses are the best.

They should be armed from head to foot, and the front rank should have Polish lances hung by a slender strap to the pommel of the saddle. They should have good, stiff swords four feet long, with triangular blades, carbines, but no pistol, as they will only increase the weight. They need stirrups, but instead of saddles, the bows only, with a pair of panels stuffed and covered with black sheepskins, which will serve as a case and to come across the horse's chest.

As for dragoons, of which twice as many are needed, the regiments should be of the same number and similarly organized. Their horses should not be over fourteen hands high nor under thirteen hands two inches. Their drill must be full of spirit and speed. They should also know the infantry drill perfectly, in case they are obliged to fight dismounted. Their arms should be the musket and the sword. Their lances should serve as pikes when

Fusileer

they are dismounted. Their saddles and harness should be the same as that of the cavalry. The men must be small and their height from five feet to five feet one inch, not over two. They should form by squadrons three deep, the same as the cavalry, and march in the same manner.

The rear rank must be taught to vault and skirmish, rallying in the intervals between the squadrons. The front and center ranks should have their muskets slung. These dragoons should do all the service of the army, cover the camp, form escorts, furnish reconnaissance elements, and discover the enemy wherever he may be.

There in general is what concerns the cavalry. It is appropriate to enter into greater detail.

Cavalry Armor

I do not know why armor has been laid aside, for nothing is both so useful or ornamental. It may be said that the invention of gun powder abolished it. It is not that; for, in the time of Henry IV and since to the year 1667, it has been worn. Powder was introduced long before.

It is certain that a naked squadron, such as ours, will stand a poor chance opposed to one armed from head to foot (assuming the numbers are equal), for what can our men to pierce them? Their only resource is to fire. It is a great advantage to reduce enemy cavalry to the necessity of firing.

This idea merits examination. I have invented a suit of armor, consisting of thin iron plates fixed on stout leather. This armor is not expensive, and the entire weight is not over thirty-five pounds. I have tried it; it is proof against a sword. I do not allege it to be the same against a bullet, especially one fired point-blank. But it will resist any that have not been well rammed, have become loose in the barrel by the movement of the horses, or come from an oblique direction.

But leave fire at that. The fire of the cavalry is not of any importance; I have always heard it said that those who have fired were beaten. If this is true, we should train them to fire. The easiest method is to give your cavalry armor such as I propose; that will make them safe from the sword and an enemy will be forced to fire. But what will happen if he fires? As soon as cavalry shall have received the fire, they will throw themselves on the enemy with irresistible *elan,* since they have nothing more to fear and wish to avenge the dangers they have just escaped.

And how can those who are unarmored, in effect, be able to defend themselves against others who are invulnerable? If the latter bestir themselves I defy anyone to kill them. If there were only two such regiments in a whole army, and they had routed a few enemy squadrons, fear and terror would spread throughout because all horsemen would appear to be armored.

I shall be answered: "The enemy will do the same thing." This is a proof that what I propose is good, since the enemy can find no other remedy except to imitate me. But this will not occur in the following campaign. We shall allow ourselves to be defeated for ten years, and perhaps for a hundred, before making a change. Whether it results from pride, laziness, or stupidity, all nations change their customs reluctantly. Even good institutions are not adopted, or only after infinite time, although often every one is convinced of their utility. In spite of all this, they are abandoned frequently to follow custom and routine. And we are told coldly: " 'Tis contrary to custom."

Discipline Wins Battles

To demonstrate what I have advanced, one need only recall the number of years during which the Gauls were always overcome by the Romans, without ever attempting to change their discipline or manner of fighting. The

Elite Corps, Louis XV

Turks are now in the same case; it is neither courage, numbers, nor riches which they lack, but discipline, order and manner of combat.

At the battle of Peterwaradin [Petrovanodin, northeastern Yugoslavia, 1716] they had more than one hundred thousand men; we with only forty thousand, defeated them. At Belgrade, they had more than two hundred thousand men; we had less than thirty thousand, and they were defeated. And this will always be the case as long as customs that are prejudicial are adhered to. These examples should persuade us never to be prejudiced in anything.

Any objections which may be made against this armor on the pretext that a shot through it will be more dangerous are false. A ball will only pierce the metal without carrying the broken pieces along with it. But even so, if the advantages of armor are weighed justly against the disadvantages, it will be found that the balance favors it infinitely. For of what consequence are a few men who die of wounds because of their armor, if battles are won and the enemy beaten? If it be considered how many troopers perish by the sword and how many are dangerously wounded by random and spent shots, accidents against which armor guarantees protection, one cannot avoid acknowledging the benefits of it.

It is nothing but indolence and relaxation of discipline that caused it to be laid aside. It is wearisome to carry a cuirass and trail a pike for months and use it a single day. But as soon as discipline is neglected in a nation, as soon as ease becomes an aim, it needs no inspiration to foretell that its ruin is near.

The Romans conquered all peoples by their discipline. In the measure that it became corrupted their success decreased. When the emperor Gratian permitted the legions to quit their cuirasses and helmets, because the soldiers complained that they were too heavy, all was lost. The barbarians whom they had defeated during so many centuries vanquished them in turn.

XI

ARMS AND EQUIPMENT FOR CAVALRY

THE men are to have rifled carbines, which carry much farther than any others and are more easily loaded, since ramming the charge is avoided; this is a very difficult feat on horseback. They must always be slung over their shoulders in an engagement, as well as on a march. Otherwise the troopers are never ready and do not keep ranks.

The swords should be slung the same as the carbines because in that position they will be less inconvenient and more ornamental. Their sabers should be three-square to keep them from cutting with them, a method of small effect. They are lighter and much stiffer than the flat kind. They should be four feet long, for a long sword is as necessary on horseback just as a short one is on foot.

I do not care for pistols because they never are effective, they are expensive, and are a useless weight and encumbrance. The front rank are to be furnished with Polish lances. These lances extend ten feet beyond the front rank, and the horses of the enemy squadrons will be terrified at the waving of the taffetas [streamers] when they are lowered. Besides, the point is not adorned. Montecuculli states in his *Memoires* that the lance is the best of all arms for the cavalry and their shock cannot be resisted. But it is essential that they be armed from head to foot.

The troopers should have a goat-skin bottle like those used in hot countries, instead of a canteen or small keg, to hold liquors in. This, with his linen, stockings, cap, a cord, and his few other necessaries, is to be put into the bottom of his sack, to roll up with his coat and fastened with two straps behind him. This will reduce that monstrous load which is now carried by the cavalry.

It is necessary from time to time to inspect the baggage

and force the men to throw away useless items. I have frequently done it. One can hardly imagine all the trash they carry with them year after year. The poor horse has to carry everything. It is no exaggeration to say that I have filled twenty wagons with rubbish which I have found in the review of a single regiment.

XII

ORGANIZATION OF CAVALRY

THE regiments of cavalry and dragoons, like those of infantry, should be composed of four centuries, each of 130 men. This will form four squadrons and the staff as in the infantry.

The squadrons of cavalry should never be reduced, nor augmented, because it takes ten years to make a cavalryman. None but veteran horses are good in war, and the cavalry should be a dependable body.

With regard to dragoons, they may be decreased or dismounted in time of peace. Provided they remain organized like infantry, they will be useful.

Extreme attention should be given, when marching by twos, that cavalry and dragoons do not drop into single file to pass a few bad spots on the road. If one does it, all will, with the result that instead of arriving in camp in six hours, it will take twelve.

A single bad defile on a march will cause delay unless the officers give it particular attention. If there are several, it will throw a whole column into disorder. In defiles you will find some halted and others galloping to overtake the leaders. Nothing is so destructive to the cavalry as this lack of discipline. It should be punished with the utmost severity. When there are holes in the road which cannot be avoided, it is much better to make a general halt and repair them than to disregard them, or else another road should be taken.

520 / Regt

Discouraging Halts

In passing through water, the horses must never be allowed to drink. A man who halts to water his horse will stop a whole army. When this happens, the officers should hasten to the spot, and, instead of fruitless reprimands and ill-timed mercy, they should instantly chastise the offender. Nothing is of such importance for the preservation of the formation of cavalry. Otherwise the affection the men have for their horses will have them halting little or much, and then it is impossible for them to recover their ranks without galloping.

Let no one think this does not make any difference. You will reach your camp at night, when you should have arrived by noon. If this is not prevented by extraordinary care and attention, a few days' march will ruin the best cavalry.

When the cavalry are to charge, it cannot be sufficiently emphasized that they must keep together and not disperse. Their standards should be sacred. Whatever happens in combat, their duty is always to rally to them. With these principles, if you can inculcate them, your cavalry will be invincible.

In charging, they first should move off at a gentle trot for a distance of about one hundred paces and increase their speed in proportion as they advance. They should not close boot to boot until they come within about twenty or thirty paces of the enemy, and this should be done at the command from an officer: "Follow me!" It is necessary to train the cavalry to this, and this movement should be like lightening. But they must be familiarized with it by constant exercise.

Charging at Full Speed

It is necessary to teach them, while in winter quarters, to gallop long distances without breaking squadron formation. A squadron that cannot charge two thousand paces at full speed without breaking is unfit for war. It is the fundamental point. When they know this, they are good,

King's Regiment

and everything else will be easy. That is everything they need to know.

The dragoons should know the same thing and, in addition, should be taught to skirmish. Their rear rank should make sorties in open ranks, return, and form again with celerity. They should be trained to fire on horseback and should know all the infantry drills.

In time of peace, and in winter quarters in time of war, their horses should be kept in condition by violent exercise, or runs, at least three times a week. The same severe usage is also proper for the heavy cavalry at those times. It is only in the field that they must be managed carefully, to keep them in flesh and the squadrons complete and strong.

The best chance of teaching them to stand fire is when the infantry is practicing. They should advance on the fire at a walk and be kept calm, accustoming them to go closer and closer. They should never be beaten, but stroked and encouraged. In the space of a month, they will be so accustomed to it that they will even put their nose on the muzzle of the muskets without any fright or surprise. Then they are all right. Nevertheless, they should not be allowed to approach too close, for if once they get burned you will not be able to bring them near again. This ordeal must be reserved for the day of battle.

Cavalry Detachments

The country in which war is being made will determine the usefulness, as well as success, of parties. Large detachments of cavalry seldom achieve anything useful unless by some prompt and vigorous expedition, such as intercepting a convoy, surprising a post, or supporting advanced parties of infantry which have been pushed ahead to cover your march. Then they are of immense value.

Suppose that the enemy intends to attack your rear guard or your baggage; he will not dare attempt it if you send out a large detachment the day before you march. He will hesitate to place himself between the body he wants to attack and the detachment he knows certainly to have gone out, although he may not know definitely the route it took nor where it is.

Detachments of this kind should be always strong. The commanding officers should be skillful and experienced in war, for this is one of the most difficult missions to execute, at least when the objective is not fixed. Of course, when ordered to seize or surprise a post, or to intercept a convoy, they have nothing to do except to march straight ahead and attack.

If you have good spy service you may be able to contrive ambushes in open campaign. Sometimes localities can be found that are hidden and permit unexpected attacks on bodies of troops that pass by. But this happens only rarely. Altogether, cavalry operations are exceedingly difficult, knowledge of the country is absolutely necessary and ability to comprehend the situation at a glance and an audacious spirit are everything.

Small Bodies Urged

Small cavalry detachments also are necessary, and they should be out every day. In general they need not consist of more than fifty troopers and should always avoid a fight. Their object is to get information and take a few prisoners.

If the enemy is bold and organizes large detachments to oppose yours, he should be watched until an opportunity is found to surprise him with double his force. You will then have gained superiority in the field and he will no longer dare to molest your small parties. You will be able to observe all his movements so that it will be impossible for him to take a step without your being informed. You will be secure and this will hinder and harass him greatly. Your foraging parties will be free from interference and he will be forced to use extreme precaution with his, because he is not master of the country and soon will have his troops worn out.

These are duties on which dragoons may be employed.

And, when they are trained, they will be infinitely superior to hussars because they are capable of the same rapidity of action and are more solid. But they need lots of practice and action. Large bodies of cavalry cannot catch them, and hussars cannot injure them. A troop of fifty dragoons has nothing to fear from a multitude of hussars. They always march at a trot and the hussars dare not follow them into close quarters.

After they have been taught by exercise and experience to know their own power, they will become so bold that they will always be harassing the main guards of the enemy. Officers will be developed in such operations, and the enemy will have nothing to oppose them with except patience.

XIII

COMBINED OPERATIONS

I AM convinced every unit that is not supported is a defeated organization, and that the principles which M. de Montecuculli has given in his *Memoires* are correct. He says that infantry should always be supported by cavalry and cavalry by infantry. Nevertheless we do not practice it. We place all our cavalry on the wings which are not supported by infantry. How are they supported? From four or five hundred paces! This destroys the assurance of the troops, for any man who has nothing behind him on which to retire or depend for aid is half beaten, and this is the reason that even the second line has sometimes given ground while the first was fighting. I have seen this many times and probably so have others. But no one seems to have sought the reason, which lies in the human heart.

It is for these reasons that I place small bodies of cavalry twenty-five or thirty paces in the rear of my infantry, and battalions in square formation in the interval between my two wings of cavalry, behind which it will be able to rally and stop the enemy cavalry.

King's Command

It is certain that the second line of cavalry will never fly so long as they see the square battalions in their front, and their appearance will also reassure that of the the first line. The battalions will maintain their ground because they cannot do anything else, and because they hope for prompt assistance from the cavalry, which, under the cover of their fire, will reappear in an instant, wishing to retrieve the disgrace of their defeat. Besides, these battalions will cover the flanks of your infantry, which is not an unimportant consideration.

Mixing Cavalry and Infantry

There are some who want to place small bodies of infantry between the intervals of cavalry; this is useless. The weakness of this formation will intimidate your infantry, for they feel that if the cavalry are defeated they are lost. And if the cavalry, who are also dependent on them, make a quick movement, they leave them behind and, perceiving they have lost their support, are soon in confusion. In addition, if your cavalry wing is defeated, the enemy can easily take you in the flank, and very quickly.

Others mix squadrons of cavalry with their infantry. This is of no value whatsoever, for as soon as the enemy attacks, his fire puts the cavalry in confusion, kills horses, and the cavalry gives way. This is enough to disrupt the infantry and make them do the same.

What are squadrons to do in this formation? Are they to stand fast, sword in hand, and wait the attack of the enemy's infantry, firing and advancing upon them with fixed bayonets? Or must they make the charge themselves? If they are repulsed, which will most probably be the case, they disorder their own infantry. To imagine that they will be able to find their own places again is hardly rational.

XIV

ARMY IN COLUMN

NOTWITHSTANDING the great regard I have for the Chevalier Follard, and that I think highly of his writings, I cannot agree with his opinion about the column. This idea seduced me at first; it looks dangerous to the enemy, but the execution of it reversed my opinion. It is necessary to analyze it to show its faults.

The Chevalier deceives himself in imagining that this column can be moved with ease. It is the heaviest body I know of, especially when it is twenty-four deep. If it happens that the files are once disordered, either by marching, the unevenness of the ground, or the enemy's cannon, no man alive can restore order. Thus it becomes a mass of soldiers who no longer have ranks or order, and where everything is confounded.

I have frequently been surprised that the column is not used to attack the enemy on the march. It is certain that a large army always takes up three or four times more ground on the march than is necessary to form it, even though marching in several columns. If, therefore, you get information of the enemy's route and the hour at which he is to begin his march, although he is at the distance of six leagues from you, you will always arrive in time to intercept him, for his head usually arrives in the new camp before his rear has left the old.

Large Intervals Breed Confusion

It is impossible to form troops scattered over such a distance without causing large intervals and dreadful confusion. I have often seen such a movement made without the enemy having thought of profiting by the occasion. And I thought they must have been bewitched.

This subject would furnish a useful chapter, for many diverse situations produce such marches. And in how many places may not one attack without risking anything?

How frequently an army is separated on its march by bad roads, rivers, difficult passes? And how many such situations will enable you to surprise some part of it? How often do opportunities present themselves of separating it, so as to be able, although inferior, to attack one part with advantage and at the same time, by the proper placement of a small number of troops, prevent its being relieved by the other?

But all these circumstances are as various and intermediate as the situations which produce them, and nothing more is required than to keep well-informed, to acquire a knowledge of the country, and to dare. You risk nothing, for as these affairs are never decisive for you, they can be for the enemy. The heads of your columns attack as they arrive and they are supported by the troops that follow them. This results from the formation itself. And you attack regiments which have no support.

But I find I have wandered from the first principles of the military art and that it is not yet time to commence on such higher subjects.

XV

USE OF SMALL ARMS

FIRE should not be used against infantry where they can envelop you or you can envelop them. But where you are separated from an enemy by hedges, ditches, rivers, hollows and such obstacles, then it is necessary to know how to aim and execute such terrible fire that it cannot be resisted.

I recommend the breech-loading musket. It can be loaded quicker, carries farther, is more accurate, and the effect is greater. In the excitement of battle, soldiers will not be able to put cartridges in the barrel without opening them. This often happens now and makes the muskets useless. They will not be able to insert two charges because the chamber will not hold them. Consequently, muskets will not burst as they often do.

To dislodge the enemy from a position on the other side of a river, from hedges, ditches and such other places where the use of small arms is necessary, I should designate an officer or noncommissioned officer to every two files. He should advance the leader of the first a pace forward and show him where he is to direct his fire, allowing him to fire at will; that is, when he has found a target.

The soldier behind him will then pass his gun forward and the others in the same manner. The file leader will thus fire four shots in succession. It would be unusual if the second or third shot does not reach its mark. The commanding officer is close by him, watches his aim, directs him where to fire, and exhorts him not to hurry. This man is not hindered, nor crowded, nor hastened by the word of command. No one presses him; he can fire at ease and aim as long as he wishes, and he can fire four times in succession.

This file having fired, the officer withdraws it and advances the second which performs in the same fashion. Then he returns the first which has had ample time to load. This can be repeated for several hours.

This fire is the most deadly of all, and I do not think any other can resist it. It would silence that by platoons or ranks, and even if they all were Caesars I would defy them to hold for a quarter of an hour. For with my guns one can fire six rounds a minute with ease. Call it four to allow for shifting guns; every musket will have fired sixty shots in a quarter of an hour, and consequently the file leaders of a regiment of five hundred men will have fired thirty thousand, not considering the light-armed forces. In an hour this will amount to 120,000 shots, and including the fire of the light-armed troops, 140,000, all better aimed than ordinary fire.

XVI

COLORS OR STANDARDS

THE general or commander-in-chief of an army should have a standard to be carried ahead of him as a mark of his rank. This also has a purpose. Anyone searching for him will know instantly where to find him, especially in battle, and the troops seeing the standard will know that the general is observing them.

Since nothing is more useful than colors or standards in action, they should be given particular attention. In the first place, they should all be of different colors so that the legions, regiments and even centuries to which they belong may be readily distinguished in combat.

The soldiers of each century should make it an article of faith never to abandon their standard. It should be sacred to them; it should be respected; and every type of ceremony should be used to make them respected and precious. This is an essential point, for after troops are once attached to them you can count on all sorts of successes; resolution and courage will be the natural consequences of it; and if, in desperate affairs, some determined man seizes a standard, he will render the whole century as brave as himself because it will follow him.

If the standards are distinguished by their different colors, the actions of every century will be conspicuous. This will create the greatest emulation because both officers and soldiers will know that they are seen and that their countenance, conduct and behaviour are not ignored by the rest of the army. For example, the first standard of a regiment that has fled will be seen, distinguished and recognized by the generals and all the troops.

The first century that shall have forced a pass, carried an entrenchment, or made a vigorous charge, will be easily distinguished, deserve praise and gain the applause of the whole army. The men as well as officers will tell of it; in the field and in garrison their exploits will be the constant subject of conversation. The desire to imitate brave actions will be aroused by praise. And these trifles will diffuse a spirit of emulation among the troops which affects both officers and soldiers and in time will make them invincible.

The number of every century should be distinguished by the color of its standard. Every standard should have a white quarter near the staff to hold the number of the legion marked in Roman numerals. Thus the designs and colors of the standards will distinguish the centuries of every legion, and the numerals, the legions themselves.

XVII

ARTILLERY AND TRANSPORT

I NEVER would have an army composed of more than ten legions, eight regiments of cavalry, and sixteen of dragoons. This would amount to thirty-four thousand foot and twelve thousand horse, a total of forty-six thousand men. With such an army, one of a hundred thousand can be stopped if the general is clever and knows how to choose his camps. A larger army is only an embarrassment. I do not say that reserves are unnecessary, but only that the combatant part of an army ought not to exceed such a number.

M. de Turenne was always victorious with armies infinitely inferior in numbers to those of his enemies because he moved more easily and knew how to select positions such that he could not be attacked while still always keeping near the enemy.

It is sometimes impossible to find a piece of ground in a whole province that will contain a hundred thousand men in order of battle. In case of an enemy, he is almost always forced to divide, in which event I can attack one of the parts; if I defeat it, I thereby intimidate the other and soon gain superiority. In short, I am convinced that the advantages which large armies have in numbers are more than lost by the encumbrance, the diversity of operations under the conflicting conduct of different commanders, the deficiency of provisions, and many other inconveniences which are inseparable from these conditions. But this is not the subject that I am discussing here, and it is only the question of proper proportions that led to this digression.

Small Cannon Efficient

Sixteen pounders are equally as useful as twenty-four pounders to batter a breach and are much less difficult to transport. Fifty of them, together with twelve mortars and ammunition in proportion, will be sufficient for such an army as I have been describing. Boats, with all the tackle to make a bridge, twelve hinged bridges for the passage of canals and small rivers, together with the necessary equipment, also are required.

For the rest of the transport and food supplies of the army, I prefer wagons made of wood, without any iron work in them. These are used by the Russians, and also we see them coming from Franche-Comté to Paris. They can travel from one end of the world to the other without damaging the roads. One man can drive four with ease. Each is drawn by two oxen. Ten of our wagons do more harm to a road than a thousand of these.

If we would only consider the disadvantages caused by our present method of transport, we should see the utility and benefit of adopting this. How many times is food totally lacking because the wagons have not been able to get up? How often is the baggage and artillery left behind, and the army forced to make a sudden halt? A little rainy weather and a hundred or two wagons are enough to destroy a good road and make it impassable; it is repaired and a hundred more wagons make it worse than it was before; put fascines on it and in no time they will be cut to pieces by the wheels which carry such a heavy weight on two points only.

All the wagons of the army should be drawn by oxen,

both because of their even pace and their economy. They can be pastured anywhere and, if there is any shortage of them, more can be obtained from the base. In addition, they require little harness. Wherever the army halts they pasture and feed themselves.

A single man can handle four wagons, each drawn by two oxen. It would require twelve or fifteen horses to haul as much as these eight oxen. The latter do not consume the forage they haul because they are sent out to pasture while the wagoners are cutting and loading it.

If one of the oxen is injured, it is killed and eaten and another one is purchased. All of these reasons induce me to prefer oxen to horses for transport. Each one, however, should be branded so that everyone can distinguish his own in the pasture.

XVIII

MILITARY DISCIPLINE

AFTER the organization of troops, military discipline is the first matter that presents itself. It is the soul of armies. If it is not established with wisdom and maintained with unshakable resolution you will have no soldiers. Regiments and armies will be only contemptible, armed mobs, more dangerous to their own country than to the enemy.

It is a false idea that discipline, subordination, and slavish obedience debase courage. It has always been noted that it is with those armies in which the severest discipline is enforced that the greatest deeds are performed.

Many generals believe that they have done everything as soon as they have issued orders, and they order a great deal because they find many abuses. This is a false principle; proceeding in this fashion, they will never reestablish discipline in an army in which it has been lost or weakened. Few orders are best, but they should be followed up with care; negligence should be punished without partiality and without distinction of rank or birth; otherwise, you will make yourself hated. One can be exact and just, and be loved at the same time as feared. Severity must be accompanied with kindness, but this should not have the appearance of pretense, but of goodness.

Flogging Should Be Mild

Whippings need not be severe. The more moderate they are the more quickly will abuses be remedied, since all the world will join in ending them.

We have a pernicious custom in France of always punishing military offences with death. A soldier caught pillaging is hanged. The result is that no one arrests him because they do not want to cause the death of a poor devil who is only trying to live. If instead, he were only turned over to the guard to be put in chains and condemned to bread and water for one, two, or three months, or put to work at any of the labors that always have to be done in an army, and then were sent to his regiment before a battle or when the general wished, everyone would agree with this punishment and the officers of the patrols would arrest them by hundreds. Soon there would be no pillaging because everyone would join in putting it under control.

At present only the unlucky are arrested. The guard and all the world, when they see them, turn the other way. The general complains because of the outrages committed; finally the provost marshal arrests one offender and he is hanged. And the soldiers say that it is only the unlucky that loses. Does this conserve discipline? No, it only causes the death of a few men without reforming the evil.

Ah, it may be said, officers also allow them to pass their posts unnoticed. There is a remedy for this abuse. It is only necessary to question soldiers who the provost marshal has captured, make them admit what posts they have passed, and send the officers in charge of them to prison for the rest of the campaign. This soon will make them vigilant, attentive and inexorable; but, when it is a question of the death of the man, there are few officers who will not risk two or three months in prison.

German Methods Praised

There are some things of great importance for discipline to which no attention is given and which officers ridicule. They even treat those as pedants who attempt to enforce them. The French, for example, ridicule the custom of the Germans of not touching dead horses. Nevertheless, it is very prudent and very wise if not carried too far. Its purpose, in armies, is to prevent soldiers from eating the carcass which, besides its uncleanliness, is unhealthy. This does not prevent them, during sieges and in case of necessity, from killing their horses and eating them. Let us judge if the infamy that is now attached to this regulation is useful or otherwise.

The Germans are reproached for whipping; it is an established military punishment among them. If a German officer strikes or otherwise abuses a soldier, he is dismissed on the complaint of the soldier. The officer is obliged to give him satisfaction in a duel, if the soldier demands it, when he is no longer under his command, without dishonoring the officer. This obligation prevails through all the military ranks, and there are often instances of generals giving satisfaction at the point of a sword to subordinate officers after they have left the service. They are unable to refuse the challenge without dishonoring themselves.

Discipline By Flogging

The French do not hesitate to strike a soldier with their hand, but they fear to use whipping as punishment because false ideals of what constitutes personal rights have destroyed its use. Nevertheless, this type of punishment is often needed, and promptly, and is neither injurious nor dishonorable. Let us compare these different customs and judge which is best for the service and which is most consistent with personal honor.

It is the same with the discipline of officers. The

French reproach the Germans for their provosts and their chains; the latter retort by exclaiming against the prisons and ropes of the French. German officers are never confined in prison, where they may be thrown in with thieves or men about to be hanged. They have a provost in every regiment; it is always an old sergeant who is given this post as a reward for his services. With respect to chains, I have never seen them used unless a criminal affair was involved. I have seen French tied with ropes. Let one balance these methods again, and it will serve to demonstrate the absurdity of condemning customs before the causes have been examined.

After having explained my ideas about infantry and cavalry, on methods of fighting and on discipline, which are, so to speak, the base and the fundamentals of the military art, I shall now proceed to the higher branches. Perhaps few will understand me, but I write for experts and to instruct myself from their criticisms. They should not be offended by the assurance with which I deliver my opinions. They should correct them; that is the fruit I expect from this work.

XIX

DEFENSE OF PLACES

I AM always astonished that no one objects to criticism of fortifying cities. These words may seem extraordinary, and I shall justify them. First examine the utility of fortresses. They cover a country, oblige the enemy to attack them or march around them, one can retire into them with troops and place them in security, they protect supplies and, during the winter, troops, artillery, ammunition, etc., are kept in security.

If these considerations are examined, it will be found that fortresses are most advantageous when they are erected at the junction of two rivers. To invest them, when so placed, it is necessary to divide an army into three parts; the defender may defeat one of these three corps before the other two come to its aid. Before being invested, such a fortress always has two open sides, nor can the enemy completely surround the fortress in a day. He will need equipment for three bridges, and these often are hazards in themselves, due to storms and floods in the campaign season.

Besides this, in holding such a post and controlling the rivers, one is master of the country. Their course can be diverted if necessary, and they permit easy supply, formation of depots and transport of munitions and all the other stores required in war.

Lacking rivers, places can be found fortified by nature so strongly that it is almost impossible to surround them, and which can be attacked only in spots. Small expense will make them impregnable. Others can be provided with locks, providing sources of water are available, and protected by extensive inundations. Everyone will admit that such situations can be found, and that, by aiding nature with art, they can be made impregnable. Nature is infinitely stronger than the works of man; why not profit from it?

Establishment of Towns

Few cities have been founded primarily for defensive purposes. Commerce has bid to their creation, and their location was chosen by hazard. In course of time they have grown and the inhabitants have surrounded them with walls for defense against the incursions of common enemies and for protection from internal disturbances in which nations are involved. All this was dictated by reason. The citizens fortified them for their own preservation, and they have defended them.

But what could be the inducement for rulers to fortify them? It could have some appearance of reason when Christianity lived in the midst of barbarism, when one city enslaved another, and when countries were devastated. But now that war is made with more moderation, what is there to fear? A town surrounded with a strong wall and capable of holding three or four hundred men, besides the inhabitants, together with some artillery, will be as secure as if the garrison consisted of as many thousands. For I maintain that these larger bodies of troops will not defend themselves longer than the four hundred and that under the terms of surrender the lot of the citizens will not be better.

But besides, what use will the enemy make of it after he has taken it? Will he fortify it? I think not. Thus he will be content with a contribution and will march on; perhaps he will not even besiege it because he will not be able to hold it. He will never hazard leaving a small garrison in it, and still less will he leave a large one because it will not be secure.

A still stronger reason persuades me that fortified cities are hard to defend. Suppose you have stored food supplies for three months for the garrison; after it is besieged you find that they last only eight days because you have not counted on twenty, thirty, or forty thousand mouths that must be fed. These are the peasants from the country who take refuge in the city and augment the number of the citizens. The wealth of a prince would not be able to establish such magazines in all the cities that might be attacked, and to renew them every year. And if he had the philosopher's stone, still he could not do it without creating a famine in the country.

Civilians Provide Problems

I imagine some one will say: "I should expel the citizens who are unable to provide their own provisions." This would be a worse desolation than the enemy could cause, for how many are there in the cities who do not live from day to day? Besides, can you be sure that you will be besieged? And if you are, will the enemy allow this multitude to withdraw tranquilly? He will drive them back into the city. What will the governor do? Will he allow these unfortunates to die from hunger? Could he justify this conduct to his ruler? What can he do then? He will be forced to supply them with provisions and surrender in eight or fifteen days.

Suppose the garrison of a city consist of five thousand men and that there are forty thousand mouths besides that. The magazine is established for three months. But the forty-five thousand will eat in one day the supplies that would have lasted the five thousand for nine days. Thus the city cannot hold for more than eleven or twelve days. Even grant that it will last twenty days; in this case it will not even need to be attacked. It is obliged to surrender, and all the millions that have been spent in fortifying it are a useless expense.

It seems to me that what I have said should demonstrate the irremediable defects of fortified cities, and that it is more advantageous for a ruler to establish his strong points in localities aided by nature, and situated to cover the country, than to fortify cities at immense expense or to augment their fortifications. It is necessary, on the contrary, after having constructed others, to destroy the fortifications of cities down to the walls. At least no thought should be given to fortifying them further or of using the money to construct fortifications for such purposes.

Notwithstanding that what I have advanced is founded on reason, I expect hardly a single person to concur, so absolute is custom and such is its power over us. A fortified place, located as I have proposed, could hold out several months or even years, provided it can be supplied, because it is not encumbered with the civil population.

The sieges in Brabant would not have been carried on with such rapid success if the governors had not calculated the duration of their defense by that of their provisions. On this account they were as impatient as the enemy to have a breach made so that they could surrender honorably. In spite of this goodwill, I have seen several governors obliged to surrender without having had the honor of marching out through the breach.

I shall not write extensively on the manner of defending fortified places, since I do not intend in this work to deal with all the phases of war in detail. My intention simply is to expose such of my ideas as seem new to me.

I have noticed in sieges that the covered ways are crowded at night with men and a great fire of small arms is constantly made from them, but that it does little damage. This is worth nothing because it fatigues the troops to excess. The soldier who has been firing all night is worn out. His musket is out of order and he passes part of the next day cleaning and repairing it and making cartridges. This deprives him of the rest that he should take, a matter of infinite consequence and which results, if great attention is not given it, in illness and a general discouragement which even goodwill cannot resist.

However, it is towards the end of a siege when the most vigor should be shown, for it is then a question of skill. The more vigor you display the more the enemy will be discouraged. It is then that sickness spreads in his camp, that forage and provisions run short, and that everything seems to contribute to his ruin, adding to the despondency of his officers and soldiers. If, added to this, they feel that your resistance is becoming stronger than it was, and that it augments in the measure in which they expected it to diminish, they will be at a loss and give themselves up wholly to despair. That is why the best troops should be reserved for sudden blows; they should not be allowed even to stick their noses over the ramparts and, especially, should be relieved from guard duty at night. As soon as they have completed their assignment, they should be returned to their quarters, their dug-outs, or wherever they are lodged.

With regard to the fire from the covered way or from the ramparts on the enemy's working parties at night, it amounts to little more than noise.

It is much better to place some barbette batteries, toward the end of the day, either in the covered way or upon the ramparts, aline them with chalk so that they are pointed in the directions desired, and fire on that line during the night. They can be withdrawn at daybreak. This fire is more deadly than small arms because it pierces fascines and gabions and the wounds it causes are mortal, since the balls are as big as walnuts. These balls will scour the breadth of the trenches and will ricochet and roll far beyond the range of musketry. Enemy cannon cannot silence them during the night and they will kill workmen and enemy cannoneers like flies.

Twelve pieces emplaced in this fashion will require only thirty-six soldiers and twelve cannoneers, and I am confident they will do more damage than a thousand men firing all night from the covered way. During all this time your troops are at rest and the next morning are in condition to make sorties or work.

It may be objected that this will use lots of powder, but soldiers firing small arms would use more. If too much is being used it is only necessary to emplace a smaller number of pieces. It will always give you an advantage; your troops will be less fatigued and consequently freer from illness. Nothing causes so much sickness as night duties.

XX

WAR IN GENERAL

I AM taking matters as they occur to me. Thus, no one should be surprised if I quit the chapter on fortification to revert to it later. It is because I considered this digression necessary here, before entering more particularly into the details of each subject. Many persons believe that it is advantageous to take the field early. They are right when it is a question of seizing an important post; otherwise, it seems to me that there is no need to hasten and that one should remain in winter quarters longer than usual. What difference does it make if the enemy lays a few sieges? He will weaken himself by doing it, and, if you attack towards autumn with well-disciplined and well-ordered army, you will ruin him.

I have always noticed that a single campaign reduces an army by a third, at least, and sometimes by a half, and that the cavalry especially is in such a pitiable state by the end of October that they are no longer able to keep the field. I would prefer to continue in quarters until then, harass the enemy with detachments, and, towards the end of a good siege, fall on him with all my forces.

I believe that I would have a good bargain and that he soon would think of withdrawing, although this might not be easy for him when opposed to troops that are well led and complete. He probably would be forced to abandon his baggage, cannon, part of his cavalry, together with all his wagons. This loss will not facilitate his going in the field so early next year. Perhaps he even might not dare to reappear. It is the affair of a month. And then one returns to winter quarters in good order, while the enemy is ruined. At this time of the year the barns are full, it is dry everywhere, and there is little sickness.

One can even turn in another direction and subsist all winter in the enemy's country. Winter need not be feared for troops so much as is commonly believed. I have made several winter campaigns in frightful climates; the men and horses kept well. There are no illnesses to fear, fevers are not prevalent as in summer and the horses are in good condition.

Subsistence of Armies

There are situations which will permit you to place your troops in security in cantonments and with abundant food. The problem is to create these conditions. An experienced general does not live at the expense of his sovereign; on the contrary he will raise contributions for the subsistence of his army and for the ensuing campaign. Being well-lodged and warm, with an abundance of everything, the soldiers are contented and happy, and live at ease.

In order to accomplish this it is necessary to know how to collect provisions and money from afar without fatiguing the troops. Large detachments are in danger of being attacked and cut off. They do not produce much and wear out the troops. To obviate this, the best way is to send circular letters to those places from which contributions are required, threatening them that parties will be sent out at a definite time to set fire to the houses of those who do not have quittances for the tax imposed. The tax should be moderate.

Following this, intelligent officers should be selected and assigned a certain number of villages to visit. They should be sent with detachments of twenty-five or thirty men and should be ordered to march only at night. The men should be ordered to refrain from pillage on pain of death. When they have arrived in the locality and it is time to determine if the villages have paid, they should send a sergeant with two men to the chief magistrate of the village to see if he has procured the quittance. If he has not, the leader of the detachment should show himself with his troops, set fire to a single house, and threaten to return and burn more. He should neither pillage, nor take the sum demanded, nor a larger one, but march away again.

Levying Food and Money

All these detachments should be assembled at the same rendezvous before they are dismissed. There they should be searched and those who are found to have stolen the slightest thing should be hanged without mercy. If, on the contrary, they have faithfully followed orders they should be rewarded. By such means, this method of raising contributions will become familiar to the troops, and the country a hundred leagues around will bring in food and money in abundance. The troops will not be exhausted, since twenty detachments a month can carry out the duty.

These detachments will not be betrayed, whatever demand the enemy may make. And since it is a calamity that they feel and that they are unable to see until its effect are felt, their terror is increased, and no one sleeps in repose unless he has paid, regardless of any prohibitions the enemy may order.

A large body engaged in exacting contributions covers little of the country and causes trouble wherever it goes. The inhabitants hide their goods and their cattle. In this state little is collected because they know the troops cannot remain long; they expect help—oftentimes they have gone for it themselves—with the result that the troops retire in haste without having accomplished much, and some are always lost.

Even when things go best, the commander of the detachment, from fear or prudence, makes the best compromise he can with the inhabitants and only brings back worn-out troops in bad condition, some food, and little money. Such is the success that contributions ordinarily have. In place of this, in the fashion I propose, everything goes well and of itself.

Mild Methods Prevail

Since only so much a month is required to be paid, the inhabitants will aid each other. They are able to furnish just so much more because they are not bothered by the presence of troops, because they have time and because they do not see any recourse except to be burned out if they do not satisfy the demands. Finally, an immense extent of country is covered. The more distant inhabitants sell their supplies to bring in money and those closer in bring in provisions. They should always be allowed to choose.

These detachments must be very unfortunate, or badly conducted, to be discoveed. For a detachment of twenty-five men can cross a whole kingdom without being captured. They march off when they are discovered and an army cannot take them.

The last war proved the truth of what I maintain. In

1710 [during the War of the Spanish Succession, when Saxe fought with the Saxons against the French] I was attacked by a party of French between Brussels and Malines. Three days later another detachment of fifty men entered Alost, which is five leagues from Brussels, at noon and carried off my baggage. At this time there were 1500 men at the gate of the town waiting for their billets which were being assigned at the city hall. I was almost captured myself. It was dangerous to go by boat from Brussels to Antwerp without a passport because it was stopped two or three times every day.

No one dared to walk in the suburbs of Brussels, Antwerp, Louvain, or Malines, without a passport. Nevertheless, the Allies were masters of all Flanders. Lille, Tournai, Mons, Douai, Ghent, Bruges, Ostend, and all the barrier towns were in their hands. There were 140,000 troops in these garrisons, and it was the middle of winter. Nevertheless, French partisans were everywhere.

This proves completely the possibility of what I advocate and convinces me that its success will be infallible.

XXI

HOW TO CONSTRUCT FORTS

WE EXCELL the Romans in the art of fortification, but we have not reached perfection. I am not particularly wise, but the great reputations of Vauban [famous French military engineer 1633-1707] and Coehorn [Dutch military engineer, 1641-1704, inventor of a portable siege mortar, named after him] do not overwhelm me. They have fortified places at enormous expense and have not made them any stronger. The speed with which they have been captured proves it.

We have modern engineers, hardly known, who have profited from their faults and surpass them infinitely. But they only hold a mean between the defects of those gentlemen and the point of perfection toward which we should strive.

I have made the plan and the profiles of an octagon fort and have calculated the time in which 4800 men could construct it. The entire work could be built in 111¾ hours or, at ten hours a day, in eleven days and one and three-quarter hours. Although these calculations are correct, they should not be depended on practically. I made them only to obtain an approximation of practicability of my plan. If the estimate of time is doubled or tripled, no mistake can be made and the difference is not very great.

The best method of employing workmen is by dividing them into four reliefs of three hours each. Then work is continuous and all the troops are employed without being worn out and work vigorously, since a soldier who works only three hours a day cannot be overworked. The work should be accompanied by the beat of the drum in cadence. Lysander with a detachment of three thousand Spartans destroyed the port of Piraeus in Athens to the sound of the flute in six hours. We still have some remnants of this custom, and it is only a few years since galley slaves at Marseilles worked in cadence to the sound of the timbrel [resembling a tamborine].

Methods of Handling Earth

For terraced works, the earth should be shoveled from step to step, or from stage to stage. Wheelbarrows are expensive and hard to push up a slope; the gentle ramps that they require considerably lengthen the distance they must be wheeled.

A soldier can easily throw his shovel of earth from a depth of from nine to twelve feet. When this is impracticable the earth must be removed in baskets. In this case the workers are divided into two groups; one digs and loads and the other carries. The pioneers, in digging, must leave steps on which to rest the baskets while they are being filled. The baskets are emptied at the places marked. All this should be done in cadence to the sound of some instrument.

It is absolutely necessary to accustom soldiers to labor. If we examine Roman history we shall find that Republic looked on ease and indolence as most formidable enemies. The consuls prepared their legions for battle by rendering them indefatigable. Rather than have them idle, they employed them on unnecessary works. Continual exercise makes good soldiers because it qualifies them for military duties; by being habituated to pain, they insensibly learn to despise danger. The transition from fatigue to rest enervates them. They compare one state with another, and idleness, that predominant passion of mankind, gains ascendancy over them. They then murmur at every trifling inconvenience, and their souls soften in their emasculated bodies.

XXII

MOUNTAIN WARFARE

THOSE who wage war in mountains should never pass through defiles without first making themselves masters of the heights; this will prevent ambushes and they can pass in security. Otherwise there is great risk of destruction or of being forced to turn about, not without great loss. And sometimes everyone perishes without a chance of saving themselves. If the passes and the heights already are occupied, it is necessary to pretend to force them to draw the attention of the enemy, while you search for a route at some other place. This disconcerts the enemy, for he has not counted on it. He does not know what disposition to make because he fears for himself and often he abandons everything. However rough and impracticable mountains may appear, passages can be found by searching for them.

Men who live in the mountains do not know them themselves because they are never obliged to seek them, and the inhabitants never should be believed for they only know their country by tradition. I have often proved the ignorance and imposture of their recitals. In such a case, it is necessary to search and see for oneself or use men who are not frightened by difficulties. You will almost always find passes when you search for them; and the enemy, who is unaware of them himself, does not know what to do and flees because he has only counted on ordinary things, which in this case are the customary roads.

Operations in Broken Country

Since the enemy, in this kind of country, is as much embarrassed as anyone can be, there is little to be feared. These are affairs of detail that decide nothing and where the more determined carries the day. There is but one thing to observe, which is to have your rear open so as to be able to send out detachments and to retire in case of need. In broken country, skill in locating artillery is of great importance. Since the enemy does not dare leave the posts he occupies, he can be fired on at will. If he abandons them, the retreat is not always successful and sometimes he can be attacked. Altogether, these affairs are never decisive and should be governed by the situation; rules of conduct cannot be prescribed. However, one should never neglect, as a course of conduct, to send out detachments in front and on the flanks of the march. These detachments of a hundred men should be supported by a second, and the second by a third, to make the cover of the main body secure.

A detachment of six hundred men can stop a whole army because along roads bordered with hedges and ditches, such as one finds in Italy and in all fertile and wet countries, they can present the same extent of front. The smallest house becomes a fortification and causes an obstinate combat. This allows time to reconnoiter and make dispositions, for in this type of country one cannot be too cautious in preventing surprises.

An audacious partisan with three or four hundred men will cause you frightful disorder and will even attack an army. If he cuts off the baggage at the beginning of darkness, he will be able to carry away a large part of it without risking much because he retreats between two obstacles and closes the gap behind him. In case he is hard pressed he marches beside the wagons and at the first house he finds he throws you back on your haunches. While this is going on the baggage he has taken from you is moving on.

If he uses this stratagem on your cavalry he will cause you frightful disorder. It is for this reason that advanced parties should cover all the avenues of your march. And they must not be too weak, for it is not a question of being informed, but of fighting to the death. Otherwise, ruin and disgrace may be your fate if you are opposed by an enemy general of understanding. He will have no difficulty in finding men in his army ready to undertake any enterprise and who see possibilities as they really are.

XXIII

RIVER CROSSINGS, SIEGES

IT IS far from being as easy as one might think to prevent the enemy from crossing a river. He can do it more easily when he is advancing to attack you than when he is retreating in front of you. In the first of these cases he presents his head to you and supports it with good dispositions and strong artillery fire. In the second he exposes his rear, which is not always so easy to withdraw, especially if he is being pressed. In addition, dispositions are not taken with such care and are not executed with as much attention for a withdrawal as for an attack, and everyone becomes negligent in the latter from a sort of timidity which results in a half defeat before the rear action starts. It would be difficult to furnish a good reason for this; it should be sought for in the natural failings of the human heart.

There is still another sort of river crossing, in which a flank is presented to the enemy. Before the battle of Turin, Prince Eugene crossed three rivers in this fashion in two days in the presence of the Duke of Orleans. The terrain between the two armies was flat, and it was a good occasion to fight, even with inferior numbers. Nothing was done, however, and the siege of Turin had to be abandoned.

In such a case, if the siege is not raised opportunely to march against the enemy, the one who is bringing help always has the advantage because the fight is never a general one for him, while it is for the one attacked. This is because the former has all his troops assembled between two rivers, his flanks are secure and he is disposed in great depth. The latter, who is investing the fortress, is dispersed and can only guard his communications across the rivers with a small number of troops.

Prompt Action Demanded

If they are defeated the entire circle is shattered, their lines are exposed on the flanks, and they soon are routed. Hesitation in such situations brings disaster. Sometimes, also, the enemy will show himself only to alarm the besiegers and to induce them to quit their posts so that he can throw help into the fortress. It is a part of the skill of the general to distinguish the true from the false.

The surest method for the besiegers is to assemble a sufficient number of troops to oppose the enemy and to leave the remainder in the lines so that they can move and attack anything that attempts to enter the fortress. But they cannot stand with arms crossed, as if bewitched, and watch an enemy cross a river before them, with his flanks exposed unmolested. It is only necessary to choose which flank to fall on, and it appears that very little resistance will be met.

At the battle of Denain [1712] Marshal de Villars would have been lost if Prince Eugene had marched against him; he had exposed his flank and crossed a river in his presence. The Prince could never imagine that the Marshal would make any such attempt in his presence, and this was what deceived him. Marshal de Villars had masked his march very skillfully. Prince Eugene watched him and examined his dispositions until 11:00 o'clock, without comprehending. All the troops were under arms; he had only to march straight ahead and the French army was lost because its flank was exposed and a large part had already crossed the Escaut.

Time Out for Dinner Costly

At 11:00 o'clock Prince Eugene said: "I think we might as well go to dinner," and recalled the troops. He was hardly at table when Lord Albemarle sent word that the head of the French army had appeared on the other side of the river and acted as if they were preparing to attack. There was still time to march and cut off a third of the French army. But Prince Eugene only ordered a few brigades on his right to march to the entrenchments of Denain, which were four leagues off. He then went with all speed to reconnoiter in person, being still unable to believe that it was the head of the French army. At length he discovered his error and saw them forming for attack. Instantly he gave up the entrenchment as lost.

I have been told (for I was not there) that he examined the enemy for a moment and chewed his glove in vexation. Whatever his feelings, he instantly gave orders to withdraw the cavalry that was in that post.

The effects produced by this affair are hardly credible. It made a difference of more than a hundred battalions in the relative strength of the two armies. Prince Eugene was obliged to place garrisons in all the adjacent cities. The Marshal, perceiving that the Allies were no longer able to carry on a siege after they had lost their magazines, withdrew about fifty battalions from the neighboring garrisons. This strengthened his army to such a degree, in comparison to the decreased strength of the Allies, that Prince Eugene no longer dared remain in the field and was forced to move all his cannon into Quesnoy, where they were afterwards captured.

With regard to forcing a passage across rivers, I believe it is hardly possible to prevent it. A river crossing ordinarily is supported by such massive artillery fire that it is impossible to prevent an advanced force from crossing, entrenching, and throwing up works to cover the bridgehead. There is nothing to be done about it during daylight, but during darkness this work can be attacked. If the attack should take place at the time the enemy has begun his crossing, everything will be thrown into confusion. Those who have crossed over will be lost, while the rest will be turned back.

But this operation must be made with a large force. If the night is allowed to pass without action, you will find that the entire hostile army has crossed. Then it is no longer an affair of detachments, but a general engagement, a hazard which the political situation does not always warrant.

There are a large number of stratagems and ruses for the passage of rivers which each employs, depending upon whether he is more or less skillful or ingenious.

Since I am dealing with affairs of detail, I should not omit how to decoy horses. There are few partisans who understand it. The decoy is a diverting stratagem to capture the enemy's horses in a foraging party or from the pasture. You must be disguised on horseback among the foragers or the pasturers on the side toward which you propose to fly. You commence by firing a few shots. Those detailed to drive up the rear answer from the other extremity of the pasture or foraging ground.

Stampeding Horses

Then they gallop toward the side fixed for the flight, shouting and firing all the way. All the horses will start to run from this side, whether they are harnessed or picketed, and will throw their riders. No matter how great their number they can be led several leagues in this manner. They should be directed into a locality bordered by hedges or ditches where they can be stopped without noise and captured at will. This is a good trick to play on the enemy and will distress him not a little. I have seen it done; but, as all things are forgotten, no one thinks of this now.

The day of the battle of Denain, when it was over, the cavalry dismounted. The Marshal de Villars, passing along the lines, gay as always, speaking to the soldiers of a regiment on the right, said to them: "Well, my lads, we have taken them." Some of them started to shout: "Long Live the King!" and threw their hats in the air. The rest of the line took it up and started shouting, throwing their hats in the air and firing, the cavalry joining in the acclamation. This frightened the horses so that they broke loose from the men and ran away. If there had been four men in front of them, they could have led them to the enemy. This accident caused considerable damage and disorder; there were a number of men wounded and a quantity of arms lost.

XXIV

DIFFERENT SITUATIONS

NOT all the situations of nature are combined as if they were planned. There are very few that are. Almost all have their features and a skillful general will profit from them. I am speaking of ravines, depressed roads, chains of lakes, and an infinity of other terrain features, all of which aid marvelously in ruses when God has had the grace to give some common sense to a man. Sometimes these things which change the situation in question so greatly are overlooked until they are forced on your attention. Then it is too late, and you see yourself reduced to being ridiculous.

It is the nature of the French to attack. When a general is unwilling to depend on the exact discipline of the troops and the great order required in pitched battles, he has only to create the occasions to fight in detail and arrange for brigade attacks. And assuredly these occasions can be found. The courage and elan which animate this nation have never been contradicted, and, since Julius Caesar (he states it himself in his *Commentaries*), I know of no example in which they have failed to make a dent in what was presented to them. Their first shock is terrible. It is only necessary to know how to renew it by skillful dispositions, and this is the business of the general.

Nothing facilitates this so much as redoubts. You can always send fresh troops to them to counterattack the enemy when he attacks. Nothing causes the enemy so much distraction, nor renders him so fearful. For while he attacks he is always afraid of being taken in the flank. On the other hand, your troops, feeling that their retreat is assured and that the enemy will not dare follow them beyond the redoubts, attack with better spirit. It is on such occasion that you can reap the greatest advantages from their vigor and impetuousness so well-known and feared among all nations and at all times. But to put them behind entrenchments is to deprive them of the means of conquering; they then are only ordinary men.

What would have been the result at Malplaquet if Marshal de Villars had taken the larger part of his army and attacked half of that of the Allies who had had the kindness to dispose themselves so that they were separated by woods and were unable to communicate with each other? The flanks and rear of the French army would have been secure.

There is more skill than one might think in making poor dispositions intentionally, but one must be able to change them into good ones in an instant. Nothing is more disconcerting to the enemy; he has counted on a certain thing, has disposed himself accordingly, and, at the instant of attacking, it has changed. I repeat: nothing confuses him so greatly and leads him into more serious faults. If he does not change his dispositions he will be defeated, and if he changes them in the presence of the enemy he still will be defeated. Human spirit cannot meet it.

XXV

LINES AND ENTRENCHMENTS

I DO NOT CARE for either of these works. When I hear talk of lines I always think I am hearing talk of the walls of China. The good ones are those that nature has made, and the good entrenchments are good dispositions and brave soldiers. I practically never have heard of entrenchments having been assaulted that were not carried. I have stated the reasons in other places. If you are inferior in numbers to the enemy, you will not be able to hold in entrenchments when they are attacked with all his forces in two or three different places at once. If you are equal with him the same will be the case. Why then go to the trouble of constructing them? They are only good for circumvallations and to prevent the enemy from throwing help into a besieged place.

The enemy's certainty that you will never dare leave them renders him audacious. He feints in your front and takes chances with flank movements that he would not dare if you were not entrenched. This boldness is infused into officers and soldiers equally. Man always fears the consequences of danger more than danger itself. I could give a multitude of examples.

Suppose that a column attacks an entrenchment and that the head reaches the edge of the ditch: if a handful of men appeared a hundred paces outside the entrenchment it is certain that the head of the column would halt or would not be followed by the rear elements. Why? The reason must be found in the human heart. Let ten men get footing on the entrenchment and all who are behind it will fly and entire battalions will abandon it. They no sooner see a troop of cavalry a half league from them than they give themselves up to flight.

When one is obliged to defend entrenchments, one should post all the battalions directly behind the parapet because, if once the enemy sets foot upon that, those in the rear will think of nothing but to save themselves. This is because of the consternation in men when something happens that they have not expected. This is a general rule in war and decides all battles and all actions. It comes from the human heart and is what induced me to compose this work. I do not believe that anyone yet has attempted to find there the reasons for the poor success of armies.

Thus when you have stationed your troops behind a parapet they hope, by their fire, to prevent the enemy from passing the ditch and mounting it. If this happens, in spite of the fire, they give themselves up for lost, lose their heads, and fly. It would be much better to post a single rank there, armed with pikes, whose business will be to push the assailant back as fast as they attempt to mount. And certainly they will execute this duty because it is what they expect and will be prepared for.

If with this, you post infantry formed according to my method into centuries at a distance of thirty paces from the entrenchment, these troops will see that they are placed there to charge the enemy as fast as he enters and attempts to form. They will not be astonished to see the enemy enter, because they expect it, and will charge vigorously. If, instead, they had been placed on the parapet, they would have fled. That is how a trifle changes everything in war and how human weaknesses cannot be managed except by allowing for them.

If the enemy endeavors to occupy the berm of the entrenchment, as happens frequently, to dislodge you from the sill, you can await him with your pikes and throw him, man by man, into the ditch as fast as they uncover themselves. And finally, if the enemy enters the entrenchment and commences to form, you charge him in detail by century. The centuries will not be surprised to see the enemy enter because they expect it and will charge vigorously because that is what they have prepared themselves to do. That is all that can be said concerning the defense of entrenchments.

Mobile Reserves Needed

In major engagements one should always have different reserves to move to the point where one sees the enemy direct the most men, a matter that is not always easy to accompish. If the enemy is skillful, you will see nothing. Thus it is necessary to place reserves as advantageously as possible, either inside or outside the entrenchments, according to terrain.

You need not fear being attacked in places where the ground is level for any considerable distance, since the enemy will not want to uncover his real purpose by exposing the main body of his force. In such places he will only have a battalion in depth. But whenever there happens to be a hill, valley, or least thing to cover his approach, there you may expect him to make all his efforts because he will hope to conceal his maneuver and numbers.

If you are able to contrive some passages in your entrenchments for a party or two to sally out at the moment when the head of some one of the enemy's columns reaches the brink of the ditch, it will be stopped infallibly, even though they may have forced the entrenchment and some have entered it. This is because they are faced with the unexpected and fear for their flanks and rear. In all probability they will take to flight without exactly knowing why.

Here are two examples which support my ideas. Caesar, wishing to relieve Amiens when it was besieged by the Gauls, arrived with his army, which was only seven thousand strong, along the banks of a stream where he entrenched himself with such haste that the barbarians, convinced that Caesar feared them, attacked the entrenchments. The Roman general had no intention of defending them. On the contrary, while the barbarians were filling up the ditch and gaining the parapet, he sallied out with his cohorts and surprised them so that they fled without a single one making the slightst attempt to defend himself.

At the siege of Alesia by the Romans, the Gauls, although infinitely superior in numbers, marched to attack them in their lines. Caesar, instead of defending them, gave orders to his troops to make a sally and fall on the enemy on one side, while he attacked them on the other. In this he succeeded so well that the Gauls were routed with considerable loss, exclusive of twenty thousand men and their general who were taken prisoners.

Undesirable Formations

If one considers the manner in which I arrange my troops (checkerboard), it is easy to understand that they can maneuver more easily than in line. It is also much easier to charge in detail than in massed lines. For this order of battle is much stronger than all the others and is not subject to any confusion, which cannot be said of the formation by battalions.

What good are several battalions drawn up four deep, one behind the other? They are unwieldy, a trifle embarrasses them, the ground, the doubling, or any other such circumstance. And if the first is repulsed, it falls in disorder on the second and throws it in confusion. But suppose that the second does not break; nevertheless it will require a long period of time before it can attack because the first, which is not broken, must be allowed to move clear of its front. And unless the enemy is so complaisant as to wait with arms folded during this time, he will certainly drive the first battalion back on the second, and the second on the third. For after having repulsed the first, he has nothing to do but to advance forward briskly. If there were thirty, one in rear of the other, he would throw them all into confusion. Yet this is what is called attacking in column by battalions! What wretched work!

My formation is decidedly different. For even if the

first battalion should be driven dack, the one which follows it can charge instantly, thus returning blow for blow. I am formed eight deep and have no fear of confusion; my charge is violent and my march rapid; I do not fear confusion and I shall always outflank the enemy, although equal in numbers. The enemy's battalions cannot remedy this fault because they do not know how to extend. Nothing certainly can be more wretched and absurd than the formations with which we fight, and I cannot conceive why the generals have not thought of changing it.

What I propose is not a novelty. It is the Roman formation. With this formation they vanquished all the nations on earth. I shall be told: "But the Romans did not have powder." It is true, but their enemies had missile weapons which had almost the same effect. The Greeks were very skillful in the art of war and well disciplined. But their large phalanx was never able to contend with the small bodies of the Romans disposed in this formation. Thus Polybius [Greek historian 204-122 B. C.] gives preference to the Roman formation. What could our battalions, which have neither body nor spirit, do against this same formation?

No matter how the centuries are posted—in a plain or in rough ground—make them sally out of a narrow pass or any other place, and you will see with what surprising speed they will form. They can be ordered to run at full speed to seize a defile, hedge, or height, and by the time the standards have arrived they will be aligned and formed. This is impossible with our battalions; for to form as they should, they need ground made to order and considerable time to make several movements. And all this is pitiful to see and often has given me a nightmare.

XXVI

OBSERVATIONS OF POLYBIUS

I HAD NOT READ Polybius throughout in 1732 when I wrote this work on war, and it was not until this year, 1740, that I had completed him. Here is what I found on the Grecian phalanx and on the Roman formation for combat. I am flattered to have thought like this contemporary of Scipio, Hannibal, and Philip [King of Macedonia, 382-336 B. C. Father of Alexander the Great.] who during several of the wars of these great men was in different armies and had distinguished commands for several years. Such an illustrious author can only justify my ideas. I leave to the readers of this work to judge if I thought like him. It is Polybius who speaks:

Having left an assurance with my readers, in the sixth book of this work, that I would choose some proper time to compare the arms and orders of battle of the Macedonians and the Romans, and to show in what respects they severally have the advantage, or are inferior to each other, I shall here take the occasion which the action now described has offered, and shall endeavor to discharge my promise.

Why Romans Won

For as the order of battle of the Macedonian armies was found, in the experience of former ages, to be superior to that of the Asiatics and the Greeks, and the Roman order of battle in the same manner surpassed that of the Africans and all the western parts of Europe; and as, in later time, these two orders have been often set in opposition to each other, it must be useful, as well as interesting, to trace out the differences between them and to explain the advantages that turned the victory to the side of the Romans in these engagements.

From such a view, instead of having recourse to chance and blindly applauding, like men of superficial understanding, the good fortune of the conquerors, we shall be able to remark with certainty the true causes of their success, and to ground our admiration upon the principles of sound sense and reason.

With regard to the battles that were fought by Hannibal and the victories which he obtained against the Romans, there is no need, on this occasion, to enter into a long discussion of them. For it was not his arms, nor his order of battle, which rendered that general superior to the Romans, but his dexterity alone and his admirable skill. In the accounts that were given by us of those engagements, we have very clearly shown that this was the cause of his success. And this remark is still more strongly confirmed, in the first place, by the final issue of the war.

For, as soon as the Romans had obtained a general whose ability was equal to that of Hannibal, they immediately became the conquerors. Add to this that Hannibal himself rejected the armor which he first had used and, having furnished the African troops with the arms that were taken from the Romans in the first battle, used afterwards no other. In the same manner also, Pyrrhus employed not only the arms, but the troops of Italy, and ranged in alternate order a company of those troops and a cohort disposed in the manner of the phalanx in all his battles with the Romans.

And yet, even with the advantage of this precaution, he was never able to obtain any clear or decisive victory against them. It was necessary to premise these observations for the sake of preventing any objection that might be made to the truth of what we shall hereafter say. Let us now return to the comparison that was proposed.

Legions Invincible

It is easy to demonstrate by many reasons that while the phalanx retains its proper form and full power of action, no force is able to stand against it in front or support the violence of its attack. When the ranks are closed in order to engage, each soldier, as he stands with his arms, occupies a space of three feet. The spears in their ancient form were seventeen cubits long (a cubit is eighteen inches), but for the sake of rendering them more commodious in action they have since been reduced to fourteen.

Of these, four cubits are contained between the part which the soldier grasps in his hands, and the lower end of the spear behind, which serves as a counterpoise to the part that is extended before him; and the length of this last part from the body of the soldier when the spear is pushed forwards with both hands against the enemy is, in consequence, ten cubits. From hence it follows that when the phalanx is closed in its proper form, and every soldier pressed within the necessary distance with respect to the man that is before him, and on his side, the spears of the fifth rank are extended to the length of two cubits, and those of the second, third, and fourth, to a still greater lengh beyond the foremost rank.

Management of Spears

It is manifest, then, that several spears, differing each from the other in the length of two cubits, are extended before every man in the foremost rank. And when it is considered, likewise, that the phalanx is formed by sixteen in depth, it will be easy to conceive what must be the weight and violence of the entire body and how great the force of its attack. In the ranks that are behind the fifth the spears cannot reach so far as to be employed agains the enemy.

In these ranks, therefore, the soldiers, instead of extending their spears forwards, rest them upon the shoulders of the men that are before them, with their points slanting upwards; and in this manner they form a kind of rampart which covers their heads and secures them against those darts which may be carried in their flight beyond the first ranks and fall upon those that are behind. But when the whole body advances to charge the enemy even these hindmost ranks are of no small use and moment; for, as they press continually upon those that are before them, they add, by their weight alone, great force to the attack and deprive the foremost ranks of the power of drawing themselves backward or retreating. Such then is the disposition of the phalanx with regard both to the whole and to the several parts.

Arms and Tactics Compared

Let us now consider the arms and the order of battle of the Romans, so that we may see by the comparison in what respect they are different from those of the Macedonians. To each of the Roman soldiers as he stands in arms is allotted the same space of three feet; but as every soldier, in time of action, is constantly in motion, being forced to shift his shield continually so that he may cover any part of his body against which a stroke is aimed and to vary the position of his sword, so as either to push or make a falling stroke, there must be also a distance of three feet, the least that can be allowed for performing these motions with advantage, between each soldier and the man that stands next to him, both on his side and behind him.

Therefore, in charging against the phalanx each Roman soldier has two Macedonians opposite him and also has ten spears which he is forced to encounter. But it is not possible for a single man to cut down these spears with his sword before they can take effect against him. Nor is it easy, on the other hand, to force his way through them. For the men that are behind add no weight to the pressure nor any strength to the swords of those that are in the foremost rank.

It will be easy, therefore, to conceive that while the phalanx retains its proper position and strength no troops, as I have before observed, can support the attack of it in front. To what cause, then, it is to be ascribed that the Roman armies are victorious and those defeated that employ the phalanx? The cause is this: in war, the times and places of action are various and indefinite, but there is only one time and place, one fixed and determinate manner of action that is suited to the phalanx. In the case of a general action, if an enemy be forced to encounter with the phalanx in the very time and place which the latter requires, it is probable in the highest degree that the phalanx always must obtain the victory. But, if it be possible to avoid an engagement in such circumstances, and indeed it is easy to do it, there is nothing to be dreaded from this order of battle.

It is well known, and an acknowledged truth, that the phalanx requires a ground that is plain and naked and free from obstacles of every kind such as trenches, breaks, obliquities, the brows of hills, or the channels of rivers, and that any of these are sufficient to impede it and to dissolve the order in which it was formed.

On the other hand again, it must readily be allowed that if it be not altogether impossible, it is at least extremely rare to find a ground containing twenty stadia or more in its extent and free from all these obstacles. But let it, however, be supposed that such a ground may perhaps be found. If the enemy, instead of coming down on it, should lead their army through the country, plundering the cities and ravaging the lands, of what use then will be the phalanx?

As long as it remains in this convenient post, it not only has no power to succor its friends, but cannot even preserve itself from ruin. For the troops that are masters of the whole country, without resisance, will easily cut off all supplies from it. And if, in the other hand, it should relinquish its own proper ground and endeavor to engage in action, the advantage is then so great against it, that it soon becomes an easy prey to the enemy.

Receiving the Enemy

But further let it be supposed that the enemy will come down into this plain; yet if he does not bring his whole army at once to receive the attack of the phalanx, or if in the instant of the charge he withdraws a little from the action, it is easy to determine what will be the consequence from the present practice of the Romans. For now our discourse is not based on bare reasoning only, but from facts which have lately happened.

When the Romans attack the phalanx in front they never employ all their forces so as to make their line equal to that of the enemy, but lead on only a part of their troops and keep the rest of the army in reserve. Now whether the troops of the phalanx break the line that is opposed to them, or whether they are broken themselves, the formation peculiar to the phalanx is alike dissolved. If they pursue the fugitives, or if, on the other hand, they retreat and are pursued, in either case they are separated from the rest of their own body. And thus there is left some space which the reserve of the Roman army takes care to seize and then charges the remaining part of the phalanx.

But the charge is not made against the front, but in flank or in rear. Since it is easy then to avoid the conditions that are favorable to the phalanx and since those, on the contrary, that are disadvantageous to it can never be avoided, it is certain that this difference alone must carry with it a decisive weight in time of action.

To this it may be added that the troops of the phalanx also are, like others, forced to march and to camp in every kind of place, to be the first to seize the advantageous posts, to invest the enemy, and to engage in sudden actions without knowing that an enemy was near.

Roman Battle Order Best

These things all happen in war and either tend greatly to promote or sometimes wholly determine the victory. But at all such times the Macedonian order of battle either cannot be employed or is employed in a manner that is altogether useless. For the troops of the phalanx lose all their strength when they engage in separate companies or man with man.

The Roman order, on the contrary, is never attended even on such occasions with any disadvantage. Among the Romans every single soldier, when he is once armed and ready for service, is alike fitted to engage in any time or place, or on any appearance of the enemy, and preserves always the same power and the same capacity in action whether in separate companies or man to man. As the parts, therefore, in the Roman order of battle are so much better contrived for use than those in the other, so the success in action must also be greater in the one than the other.

If I have been long in examining this subject, it was because many of the Greeks, at the time when the Macedonians were defeated, regarded that event as a thing surpassing all belief, and because many others also may hereafter wish to know for what reasons and in what particular respects the order of the phalanx is excelled by the arms and the order of battle of the Romans.

XXVII

ATTACK ON ENTRENCHMENTS

WHEN an entrenchment is to be attacked an attempt should be made to extend the lines as much as possible. This will make the enemy fearful everywhere so that he will not withdraw troops from any point to reinforce that which you intend to attack, even after he discovers it. This makes many of his troops useless. To effect this, all the battalions which are used only to deceive the enemy should be drawn up four deep and march in line. The rest of the maneuver and the preparations for the real assault should take place behind them.

This is what is called masking the attack. This part of the military art depends on the imagination; a general can elaborate it as much as he pleases. Everything is good, since the certainty that he will not be attacked permits him to do whatever he thinks apropos. He can make use of ravines, valleys, hedges, and a thousand other things.

In charging by centuries there need be no fear of confusion. Each centurion will consider it a matter of honor for his standard, and among them it is impossible that there will not be some who will seek to risk their lives to distinguish themselves. This is because, according to my system, the behavior of every century becomes conspicuous by the distinction of their colors.

In approaching the entrenchment, the light-armed troops should be advanced to draw the enemy's fire. They should be supported by other troops. Finally, after the firing has commenced, the centuries should come up and charge. If they are repulsed, others should replace them before the first engaged have time to flee, and force and numbers will surmount all obstacles. The centuries which have been drawn up four deep should arrive at the same time, provided you have forced the entrenchment in several places at once. The enemy battalions which are between the two forces and who see the line advance will fly and the line will gain the parapet. After that the troops can be reformed and the enemy, during this time, will retire because he will imagine that he has done everything possible.

There is still another method of attacking entrenchments, entirely different from this, and which is equally as good. But it must be favored by the terrain and the terrain must be known perfectly.

When there are ravines or hollows near the entrenchments where troops can be concealed during the march without discovery by the enemy, several columns, with large intervals between them, should be marched towards him. He will give all his attention to these columns, dispose his troops to meet them and strip his entrenchments. When the columns attack, all the enemy will move toward them and the troops who have remained concealed can appear suddenly and attack the parts of the entrenchments that have been abandoned.

Seeing this, the troops that have been opposing the attacks of the columns will be surprised and lose their heads, since they had not expected this event. They then will leave the attack, under pretext of running to defend the entrenchment, but actually from the terror that overcomes them. Thus you will be able to enter the entrenchment at the points of the true and false attacks at the same time.

Of all the varieties of war, the defense of entrenchments seems to me to be the most difficult. And although I have indicated the methods that seem best to me of the different manners of defense, I do not have much faith in the best. As far as I am concerned, I do not believe in constructing them. Redoubts are my favorite works, and I shall speak of them next.

XXVIII

ADVANTAGES OF REDOUBTS

IT BEHOOVES ME to justify by facts the high opinion that I have of redoubts.

The arms of Charles XII, king of Sweden, had always been victorious before the battle of Pultowa. [1709, when he was defeated by Peter the Great]. Their superiority over those of the Russians is almost incredible. It was not unusual for ten or twelve thousand Swedes to force entrenchments defended by fifty, sixty, or even eighty thousand Russians, and cut them to pieces. The Swedes never inquired about the number of the Russians, but only where they were.

Czar Peter, the greatest man of his age, bore the poor success of this war with a patience equal to his genius and did not stop fighting in order to give his troops experience.

In the course of his adversities, the king of Sweden laid siege to Pultowa. The Czar held a council of war at which various opinions were expressed. Some were for surrounding the king of Sweden with the Russian army and for throwing up a large entrenchment to force him to surrender. Other generals were for devastating, the country within a hundred leagues to reduce him by famine. This advice was not, in my opinion, at all bad, and the Czar inclined toward it. Other generals said that this expedient could always be adopted, but that they first should hazard a battle, since Pultowa and its garrison were in danger of being captured by the invincible determination of the King of Sweden. Here he would find a large depot and all the supplies to subsist him to cross the desert it was proposed to create around him. This resolution was adopted. Then the Czar addressed them and said:

Peter Explains His Plan

"Since we have decided to fight the King of Sweden, we should agree on the method and choose the best. The Swedes are impetuous, well disciplined, trained, and skillful. Our troops do not lack in resolution, but they do not have these advantages. It is necessary therefore, to counteract the Swedish superiority. They often have forced our entrenchments, and in the open our troops have always been defeated by the skill and facility with which they maneuver. Hence it is necessary to break up their maneuver and render it futile.

"To accomplish this, I should march toward the Swedish army, throw up several redoubts with deep ditches along the front of our infantry, garrison them with infantry and protect them with palisades and brush. This will only require a few hours labor, and we shall await the enemy behind these redoubts. He will have to raise the siege to attack us. When he has lost a lot of men and is weakened and in disorder, we shall attack. There can be no doubt that he will raise the siege and come to attack us as soon as he sees us approahing him. It is necessary, therefore, to arrange our march so that we will reach his vicinity toward the end of the day, so that he will withhold his attack until morning. During the night we will raise the redoubts."

Thus spoke the sovereign of the Russias, and all the council approved his dispositions. The orders were given for the march, the tools, the cannon, the fascines, the palisades, etc. Late in the afternoon, July 8, 1709, the Czar arrived in the vicinity of the king of Sweden.

The king of Sweden, although wounded, informed his generals that he intended to attack the Russian army in the morning. They made the necessary dispositions, drew up the troops, and marched a little before daybreak.

Construction of Redoubts

The Czar had thrown up seven redoubts in his front. They were constructed with care, and each was manned with two battalions of infantry. They were supplied with everything needed for their defense. All the Russian infantry was behind them, and the cavalry was on the wings. Thus it was impossible to reach the Russian infantry without taking the redoubts, since they could not be left intact in the rear and it was impossible to pass between them without being destroyed by their fire.

Neither the king of Sweden, nor his generals, were aware of these dispositions until they ran into them. But, since the machine had been put in motion, it was now impossible to stop it. The two wings of the Swedish cavalry routed that of the Russians and pursued them too far. The center was stopped by the redoubts. The Swedes attacked them and were resisted obstinately.

Every soldier knows the difficulty of taking a good redoubt. It requires a special formation with several battalions and fifteen or twenty companies of grenadiers, in order to attack on several sides at the same time, and even then success is uncertain. Nevertheless the Swedes carried three of these and were repulsed with great loss at the others. It was inevitable that all the Swedish infantry was thrown into disorder in attacking the redoubts, while that of the Russians, drawn up in order at the distance of two hundred paces, watched the scene in tranquility.

The King and the Swedish generals saw the danger in which they were involved, but the inaction of the Russians gave them some hope that they might be able to withdraw. It was absolutely impossible to do it in good order since everything was disorganized, attacks were being made uselessly, or men stood and were killed. To retire was the only step that could be taken. The troops that had taken the redoubts were withdrawn and the rest followed.

Instant Attack Urged

There was no way to form within range of the fire of the redoubts and, consequently, everyone was withdrawn, disorganized and intermingled. In the meantime the Czar called his generals and asked their advice. M. Allart, one of the juniors, without giving anyone else a chance to say a word, spoke to his master as follows: "If Your Majesty does not attack the Swedes instantly, there will be no time afterwards." Without delay the whole line moved forward in good order, pikes high, in the intervals between the redoubts, which were left garrisoned to protect the retreat in case it became necessary.

The Swedes had hardly halted to reform and restore order when they saw the Russians at their heels. They again became disorganized and the confusion was general. Nevertheless they did not fly. They even made a brave effort to turn as if to charge. But order, which is the soul of battle, was lacking, and they were dispersed without resistance.

The Russians, being unaccustomed to conquer, did not

dare follow them, and the Swedes withdrew without interference to Boristhene, where they all were later taken prisoner. That is how to render fortune favorable by skillful dispositions.

If the Russians, who at this time were inexperienced and discouraged by a series of misfortunes, were able to conquer with these dispositions, what success may not be expected from them with a brave and spirited nation whose instinct is to attack? For, although one is on the defensive, all the advantages of attacking are conserved. The enemy is charged with brigades that are advanced in the measure that the enemy attacks some one of the redoubts.

Time and Opportunity

The charge is frequently renewed and always with fresh troops. They await the order with impatience and attack vigorously because they are seen and supported, and especially because they know their retreat is secure. Panic, which sometimes seizes armies, need not be feared, and you make yourself master of the favorable moment which is capable of deciding the combat. By that I mean the time at which the enemy is in disorder. What an advantage to be able to await this moment with assurance!

The Russians did not profit from all the advantages that their dispositions afforded them. They allowed three of their redoubts to be taken before their faces, without attempting to aid them. This would discourage the defenders of the others, intimidate their own troops and augment the audacity of the Swedes. One may, therefore, venture to say that it was the dispositions alone which conquered the Swedes in this action, without the Russian troops having contributed greatly to the victory.

These redoubts are also the more advantageous in that they require but little time for their construction and are useful in an infinity of situations. A single one is frequently sufficient to stop a whole army in a terrain corridor. They can be used to prevent your being harassed on a critical march, to support one of your wings, to divide a piece of ground, to occupy a large space when there are not enough troops to support a flank on a wood, a marsh, a river, etc.

XXIX

SPIES AND GUIDES

TOO much attention cannot be paid to spies and guides. Montecuculli says that they are like eyes and are equally necessary to a general. He is right. Too much money cannot be spent to get good ones. These men should be chosen in the country where the war is being fought. They should be intelligent, cunning and discreet. They should be placed everywhere, among the officers, the generals, the sutlers and especially among purveyors of provisions, because their stores, magazines and other preparations furnish the best intelligence concerning the real designs of the enemy.

Spies should not know one another. There should be several ranks of them. Some should associate with soldiers; others should follow the army under the guise of peddlers. These should know one of their companions of first rank from whom they receive anything that is to be conveyed to the general who pays them. This detail should be committed to one who is faithful and intelligent. He should report his activities every day, and it should be certain that he is incorruptible.

XXX

SIGNS TO BE WATCHED

THERE are particular signs in war that must be studied and by which judgments can be formed with some certainty. The knowledge you have of the enemy and his customs will contribute a great deal to this. And there are signs common to all nations.

During a siege, for example, you discover towards the horizon and on the heights, as evening approaches, unorganized and idle groups of men looking at the city; this is a sure sign that a considerable attack is being prepared. This is because attacks are made with elements of different corps and thus are known to the whole army. Those who are not to take part in the attack gather on the heights to watch it at their ease.

When your camp is near that of the enemy and you hear much firing in it, you may expect an engagement the day following because the men are discharging and cleaning their arms.

When you are in the presence of the enemy under arms and you see the soldiers changing shirts, it is certain that you are going to be attacked, because they put on all their shirts, one over the other, in order not to lose any.

If there is any extensive movement in the enemy's army, this can be judged from several leagues by the dust which is never raised except for several reasons. The dust caused by foraging columns is not the same as that of columns on the march, but you should be able to distinguish the difference.

You also can judge the direction of the enemy's movement by the reflection of the sun on his arms. If the rays are perpendicular, he marches towards you; if they are varied and infrequent, he retreats; if they slant from right to left, he is moving towards the left; if, in the contrary, they slant from left to right, his march is to the right. If there is a great amount of dust in his camp, not raised by foraging parties, he is sending off his sutlers and baggage, and you can be certain that he will march soon.

Attack on Marching Enemy

This will give you time to make your dispositions to attack him on the march. You should know if it is practicable for him to march in your direction, whether that is his intention, and what way it is most probable that he will march. This can be judged from his position, his supplies, his depots, the terrain and, in short, his conduct in general.

Sometimes he places his ovens on his right or on his left. If you know the time and the quantity of his baking and if you are covered by a small stream, you can make a flank movement with your whole army. If he imitates you, as sometimes he is forced to do, you can return suddenly and attack the ovens with ten or twelve thousand men. The expedition should be accomplished before he is aware of it because you always have several hours advantage of him, exclusive of the time that may elapse between his receipt of intelligence and the confirmation of it. He will undoubtedly wait before he puts his army in motion, so that in all probability he will receive information of the attack of his depot before he has given orders for the march.

There are an infinite number of such stratagems in war that can be employed with little risk. Their consequences are often as great as those of a complete victory. They may force the enemy to attack you at a disadvantage or even to retreat shamefully with a superior army. And you will have risked little or nothing.

XXXI

THE GENERAL COMMANDING

I HAVE formed a picture of a general commanding which is not chimerical. I have seen such men.

The first of all qualities is COURAGE. Without this the others are of little value, since they cannot be used. The second is INTELLIGENCE, which must be strong and fertile in expedients. The third is HEALTH.

He should possess a talent for sudden and appropriate improvisation. He should be able to penetrate the minds of other men, while remaining impenetrable himself. He should be endowed with the capacity of being prepared for everything, with activity accompanied by judgment, with skill to make a proper decision on all occasions, and with exactness of discernment.

He should have a good disposition free from caprice

and be a stranger to hatred. He should punish without mercy, especially those who are dearest to him, but never from anger. He should always be grieved when he is forced to execute the military rules and should have the example of Manlius constantly before his eyes. He should discard the idea that it is he who punishes and should persuade himself and others that he only administers the military laws. With these qualities, he will be loved, he will be feared and, without doubt, obeyed.

What a General Must Know

The functions of a general are infinite. He must know how to subsist his army and how to husband it; how to place it so that he will not be forced to fight except when he chooses; how to form his troops in an infinity of different dispositions; how to profit from that favorable moment which occurs in all battles and which decides their success. All these things are of immense importance and are as varied as the situations and dispositions which produce them.

In order to see all these things the general should be occupied with nothing else on the day of battle. The inspection of the terrain and the disposition of his troops should be prompt, like the flight of an eagle. This done, his orders should be short and simple, as for instance: "The first line will attack and the second will be in support."

The generals under his command must be incompetent indeed if they do not know how to execute this order and to perform the proper maneuvers with their respective divisions. Thus the commander in chief will not be forced to occupy himself with it nor be embarrassed with details. For if he attempts to be a battle sergeant and be everywhere himself, he will resemble the fly in the fable that thought he was driving the coach.

Thus, on the day of battle, I should want the general to do nothing. His observations will be better for it, his judgment will be more sane, and he will be in better state to profit from the situations in which the enemy finds himself during the engagement. And when he sees an occasion, he should unleash his energies, hasten to the critical point at top speed, seize the first troops available, advance them rapidly and lead them in person. These are the strokes that decide battles and gain victories. The important thing is to see the opportunity and to know how to use it.

Prince Eugene possessed this quality, which is the greatest in the art of war and which is the test of the most elevated genius. I have applied myself to the study of this great man and on this point can venture to say that I understand him.

Unskilled Commanders

Many commanding generals only spend their time on the day of battle in making their troops march in a straight line, in seeing that they keep their proper distances, in answering questions which their aides de camp come to ask, in sending them hither and thither, and in running about incessantly themselves. In short, they try to do everything and, as a result, do nothing. They appear to me like men with their heads turned, who no longer see anything and who only are able to do what they have done all their lives, which is to conduct troops methodically under the orders of a commander.

How does this happen? It is because very few men occupy themselves with the higher problems of war. They pass their lives drilling troops and believe that this is the only branch of the military act. When they arrive at the command of armies they are totally ignorant and, in default of knowing what should be done, they do only what they know.

One of the branches of the art of war, that is to say drill and the method of fighting, is methodical; the other is intellectual. For the conduct of the latter it is essential that ordinary men should not be chosen.

Unless a man is born with talent for war, he will never

be other than a mediocre general. It is the same with all talents; in painting, or in music, or in poetry, talent must be inherent for excellence. All sublime arts are alike in this respect. That is why we see so few outstanding men in any science. Centuries pass without producing one. Application rectifies ideas, but does not furnish a soul, for that is the work of nature.

Roads to Defeat

I have seen very good colonels become very bad generals. I have known others who were great takers of villages, excellent for maneuvers within an army, but who, outside of that, were not even able to lead a thousand men in war, who lost their heads completely and were unable to make any decision.

If such a man arrives at the command of an army, he will seek to save himself by his dispositions, because he has no other resources. In attempting to make them understood better he will confuse the spirit of his whole army with multitudinous messages. Since the least circumstances changes everything in war, he will want to change his arrangements, will throw everything in horrible confusion, and infallibly will be defeated.

One should, once for all, establish standard combat procedures known to the troops, as well as the general who leads them. These are general rules, such as: preserving proper distances on the march; when charging to charge vigorously; to fill up intervals in the first line from the second. No written instructions are required for this; it is the A-B-C of the troops and nothing is simpler. And the generals should not give all their attention to these matters as most of them do.

But what the general should do, is to observe the attitude of the enemy, the movements he makes, or where he directs his troops. He should endeavor, by a feint at one point, to draw his troops from another, to confuse him, to seize every opportunity, and to know how to deliver the death thrust at the proper place. But to be capable of all this, he should preserve an unfettered mind and not occupy himself with trifles.

XXXII

PITCHED BATTLES OPPOSED

I DO not favor pitched battles, especially at the beginning of a war, and I am convinced that a skillful general could make war all his life without being forced into one.

Nothing so reduces the enemy to absurdity as this method; nothing advances affairs better. Frequent small engagements will dissipate the enemy until he is forced to hide from you.

I do not mean to say by this that when an opportunity occurs to crush the enemy that he should not be attacked, nor that advantage should not be taken of his mistakes. But I do mean that war can be made without leaving anything to chance. And this is the highest point of perfection and skill in a general. But when a battle is joined under favorable circumstances, one should know how to profit from victory and, above all, should not be contented to have won a battle in accordance with the present commendable custom.

The proverb: "A bridge of gold should be made for the enemy," in connection with his retreat, is followed religiously. This is false. On the contrary, the pursuit should be pushed to the limit. And the retreat which had appeared such a satisfactory solution will be turned into a rout. A detachment of ten thousand men can destroy an army of one hundred thousand in flight. Nothing inspires so much terror or occasions so much damage, for everything is lost. Substantial efforts are required to restore the defeated army, and in addition you are rid of the enemy for a long time. But many generals do not worry about finishing the war so soon.

If I wished to cite examples to support my opinion I could find an infinite number. I shall mention but one.

As the French army, at the battle of Ramillies, [1706, one of Marlborough's famous victories over the French] was retreating in good order on a narrow plateau, bordered on both sides with deep ravines, the Allied cavalry followed it at a slow pace as if they were marching for exercise. And the French army also withdrew gently, twenty or more deep at times, on account of the narrowness of the ground. An English squadron approached two battalions of French and commenced to fire. These two battalions, believing that they were going to be attacked, faced about and fired a general discharge on the squadron. What happened? All the French troops gave way at the noise of the discharge. The cavalry fled at a gallop and the infantry threw itself into the two ravines in horrible confusion, so that in an instant the ground was clear and no one was to be seen.

Can anyone boast to me, after that, of the good order of retreats and the prudence of those who build a "bridge of gold" for the enemy after they have been defeated in battle? I should say that they serve their master badly.

This is not to say that it is necessary to give yourself up totally to the pursuit and follow the enemy with all your forces. A corps should be ordered to push as long as the day lasts and to follow in good order. Once the enemy has taken flight they can be chased with no better weapons than air-filled bladders. But if the officer you have ordered in pursuit prides himself upon the regularity of his formation and the precautions of his march, that is to say if he maneuvers like the army which he follows should, there is no use in having sent him. He must attack, push, and pursue without cease. All maneuvers are good then; it is only precautions that are worthless.